NELSON MATHS

AUSTRALIAN CURRICULUM

4

Student Book

Pauline Rogers

NELSON
A Cengage Company

Australia • Brazil • Japan • Korea • Mexico • Singapore • Spain • United Kingdom • United States

Contents

Identifying Odd and Even Numbers

Colour the odd numbers blue and the even numbers pink to complete the picture.

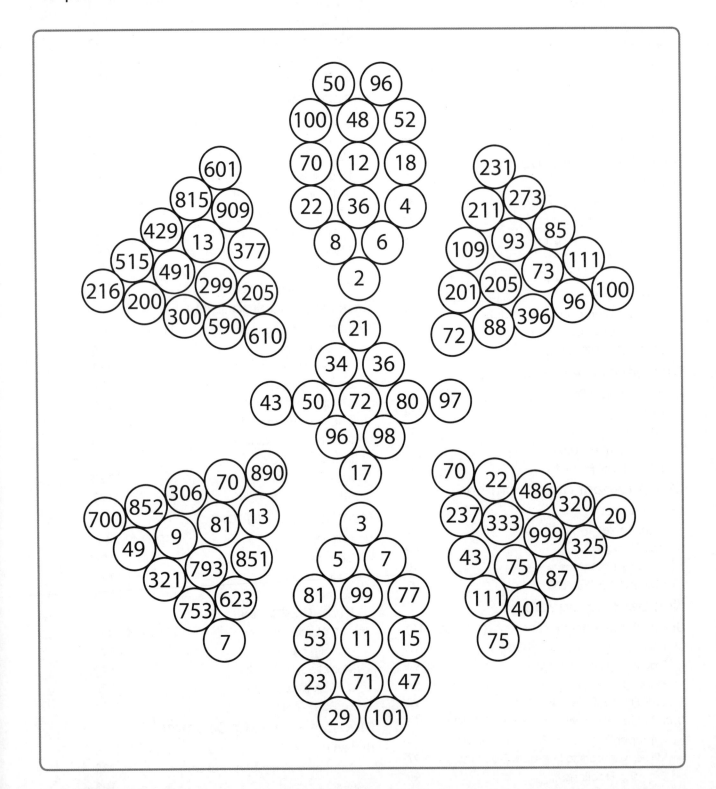

4   Unit 1   **Odd and Even Numbers** (TRB pp. 20–23)
Number and place value Investigate and use the properties of odd and even numbers
(ACMNA071) AC

Adding and Subtracting Odd and Even Numbers

1 Circle the even numbers in red and the odd numbers in blue.

    a 53   **b** 21   **c** 42   **d** 17

    e 20   **f** 36   **g** 48   **h** 51

2 Complete the equations.

    a $42 + 12 =$ **b** $34 + 14 =$ **c** $17 + 11 =$ **d** $21 + 15 =$

    e $22 + 23 =$ **f** $42 + 17 =$ **g** $35 + 13 =$ **h** $21 + 27 =$

3 Describe what you noticed about adding odd and even numbers, e.g. *When two even numbers are added, the answer is always ...*

_____

_____

4 Complete the equations.

    a $47 - 15 =$ **b** $39 - 13 =$ **c** $48 - 22 =$ **d** $56 - 34 =$

    e $47 - 12 =$ **f** $53 - 31 =$ **g** $58 - 37 =$ **h** $26 - 14 =$

5 Describe what you noticed about subtracting odd and even numbers, e.g. *When an even number is subtracted from an even number, the answer is always ...*

_____

_____

Extension: Do these rules apply with big numbers? (Use a calculator to help!) Make a prediction.

_____

_____

    a $2\,105 + 4\,583 =$ **b** $21\,117 + 41\,359 =$

    c $27\,568 - 10\,582 =$ **d** $33\,689 - 10\,274 =$

Unit 1   **Odd and Even Numbers** (TRB pp. 20–23)
Number and place value Investigate and use the properties of odd and even numbers
(ACMNA071) **AC**

5

Multiplying Odd and Even Numbers

1 Complete the number sequences with the next 5 numbers.

a 5, 10, 15, 20, _____, _____, _____, _____, _____

b 2, 4, 6, 8, _____, _____, _____, _____, _____

c 3, 6, 9, 12, 15, _____, _____, _____, _____, _____

d 6, 12, 18, 24, _____, _____, _____, _____, _____

2 What do you notice about the number sequences that start with:

a an odd number?

_____

b an even number?

_____

3 Complete the grids and statements.

×	2	4	6	8
3				

odd × even = _____

×	4	6	8	10
2				

even × even = _____

×	3	5	7	9
3				

odd × odd = _____

×	3	5	7	9
4				

even × odd = _____

4 Predict whether the answers will be odd or even. (You do not need to work out the multiplication equations.)

a 9 × 7 _____   **b** 12 × 11 _____   **c** 12 × 14 _____   **d** 13 × 13 _____

e 15 × 17 _____   **f** 18 × 20 _____   **g** 22 × 48 _____   **h** 16 × 18 _____

Extension: Do these rules apply with big numbers? (Use a calculator to help!) Make a prediction. _____

a 2 155 × 13 =   **b** 1 140 × 8 =

c 1 153 × 16 =   **d** 2 246 × 9 =

DATE:

STUDENT ASSESSMENT

1 Circle the numbers that are odd.

45   72   81   99   100   106   125   132   144   207

2 Add the numbers and state whether the answers are odd or even.

a 45 + 51 = _____   **b** 72 + 25 = _____

c 64 + 24 = _____   **d** 125 + 20 = _____

3 Predict whether the answers will be odd or even. (You do not need to work out the equations.)

a 61 – 15 = _____   **b** 97 – 78 = _____

c 102 – 78 = _____   **d** 316 – 69 = _____

4 Write 2 multiplication equations that give the following types of answers:

a even _____   _____

b odd _____   _____

5 Complete the statements.

a odd + odd = _____   **b** even + odd = _____

c odd × even = _____   **d** even × even = _____

e odd – even = _____   **f** even – odd = _____

6 Provide 2 different division equations that give even answers.

_____   _____

7 Using ideas about odd and even numbers, predict whether the answers to the equations will be odd or even.

a 4581 + 379 = _____   **b** 4026 – 987 = _____

c 45 × 63 = _____   **d** 4569 ÷ 3 = _____

8 How will your understanding of odd and even numbers help you check the accuracy of your calculations?

_____

Unit 1 **Odd and Even Numbers** (TRB pp. 20–23)
Number and place value Investigate and use the properties of odd and even numbers
(ACMNA071) **AC**

7

Matching Words and Numerals

Match the words and numerals that mean the same number.
Write the letter that corresponds to the matching words
on the line above each numeral to solve the mystery.

Why is the maths book so confused?

A seventy-two _____

B five thousand _____

C eleven thousand and fifty-five _____

E six thousand, eight hundred and fifty-one _____

F sixteen thousand, two hundred and twenty-seven _____

I two thousand and eleven _____

L ten thousand and twelve _____

M ninety thousand, five hundred and twenty-one _____

O twenty-two thousand, seven hundred and forty-two _____

P four hundred and eighty-five _____

R six thousand, one hundred and twenty-two _____

S eighty-one thousand, six hundred and seventy-seven _____

T four thousand and four _____

U forty-three thousand, one hundred and nineteen _____

5000	6851	11055	72	43119	81677	6851	2011	4004
2011	81677	16227	43119	10012	10012		22742	16227
485	6122	22742	5000	10012	6851	90521	81677

Unit 2 **Numbers to Tens of Thousands** (TRB pp. 24–27)
Number and place value Recognise, represent and order numbers to at least tens of thousands
(ACMNA072) **AC**

Ordering Numbers

1 Circle the **largest** number in each set.

    a 235, 465, 654, 352   **b** 1 256, 2 563, 3 564, 5 643

    c 1 254, 5 241, 5 412, 1 524   **d** 3 200, 2 300, 1 200, 2 500

    e 10 256, 10 562, 10 652, 10 255   **f** 12 125, 14 325, 10 265, 20 456

2 Circle the **smallest** number in each pair.

    a four hundred and fifty-one   three hundred and ninety-five

    b one thousand and fifteen   one thousand, three hundred and eleven

    c four thousand and twenty-five   forty thousand and five

    d twenty-four thousand, seven   twelve thousand, nine hundred and
       hundred and twenty-two   seventy-nine

    e sixty thousand and forty-nine   sixty thousand, seven hundred and one

    f eleven thousand, four   ten thousand and fifty-six
       hundred and thirty-two

3 Order the sets of numbers from **smallest** to **largest**.

    a   421   751   936   562   105

    _____

    b   2 152   2 264   2 863   2 083   2 598

    _____

    c   10 256   11 256   13 568   12 560   11 798

    _____

    d   21 975   12 795   17 951   15 179   20 498

    _____

4 Which number is **larger**: 23 568 or 23 658? _____

5 Which number is **smaller**: twenty-five thousand and eighty-one
    or twenty-five thousand, eight hundred and one? _____

Unit 2   **Numbers to Tens of Thousands** (TRB pp. 24–27)
Number and place value Recognise, represent and order numbers to at least tens of thousands
(ACMNA072) **AC**

9

Number Lines and 5-Digit Numbers

1 Write the letter for each number on the correct location on the number line.

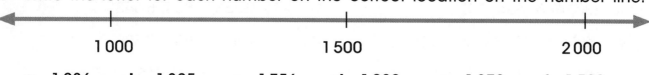

1 000   1 500   2 000

a 1 236   **b** 1 325   **c** 1 756   **d** 1 980   **e** 1 079   **f** 1 500

2 Circle the numbers that are out of place on each number line.

a

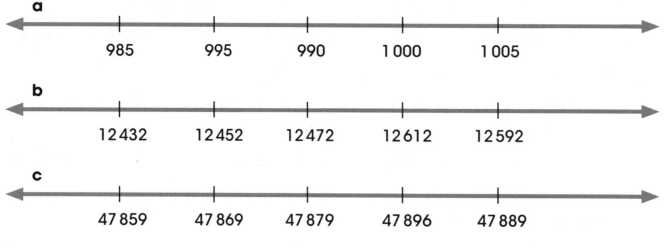

985   995   990   1 000   1 005

b

12 432   12 452   12 472   12 612   12 592

c

47 859   47 869   47 879   47 896   47 889

3 Place the numbers on the number lines.

a 335   375   320   310   350

b 3 562   3 589   3 629   3 700   3 710

c 2 156   4 729   6 987   4 689   8 678

d 48 369   49 693   48 936   47 996   47 396

4 List things that are important to remember when constructing a number line.

Unit 2 **Numbers to Tens of Thousands** (TRB pp. 24–27)
Number and place value Recognise, represent and order numbers to at least tens of thousands
(ACMNA072) **AC**

STUDENT ASSESSMENT

1 Draw a line matching the numerals and the words.

a 639   nine thousand, eight hundred and sixty-three

b 4798   seventeen thousand, three hundred and sixty-eight

c 9863   twenty-one thousand, five hundred and sixty-eight

d 10256   seventy-one thousand, six hundred and eighty-nine

e 21568   four thousand, seven hundred and ninety-eight

f 17368   six hundred and thirty-nine

g 71689   ten thousand, two hundred and fifty-six

2 Write the numbers in words.

a 56248 _____

b 32001 _____

3 Write the numbers in numerals.

a thirteen thousand, seven hundred and five _____

b forty thousand and twenty-one _____

4 Order each set of numbers from **smallest** to **largest**.

a   23568   32568   32685   23685   30256

_____

b   71895   86000   18972   85123   69999

_____

5 Complete a number line for the set of numbers.

      12035   11950   12525   11360   12000

Unit
2

Numbers to Tens of Thousands (TRB pp. 24–27)
Number and place value Recognise, represent and order numbers to at least tens of thousands
(ACMNA072) **AC**

11

Hidden Number

1 Colour the place described for each number in red. The first one is done.

2	4	7	6	3	hundreds, tens, ones
8	4	4	8	4	hundreds, ones
3	2	5	8	1	ones, hundreds
6	0	4	8	0	hundreds, ones
1	7	4	1	3	tens of thousands, thousands, hundreds, tens, ones
5	4	7	8	9	ones, tens, hundreds, tens of thousands, thousands
1	7	4	3	2	tens of thousands, hundreds, ones
3	2	2	8	7	ones, hundreds, tens of thousands
9	9	8	5	1	hundreds, tens of thousands, ones
7	4	5	2	8	tens of thousands, ones, hundreds
1	1	0	8	0	hundreds, tens of thousands, tens, ones
7	2	5	8	7	hundreds, tens of thousands, ones
1	0	3	3	3	tens of thousands, ones, hundreds
4	7	8	0	1	ones, hundreds, tens of thousands
9	7	5	6	3	thousands, tens of thousands, ones, hundreds

Turn your page sideways so the digits are at the top. What is the secret number you have shaded? _____

2 Select 3 numbers from the table. On another sheet of paper, list them and write them in words.

Spider Web

1 Join the dots between each set of 2 or 3 ways to write the same number to make a spider web. The first one has been done.

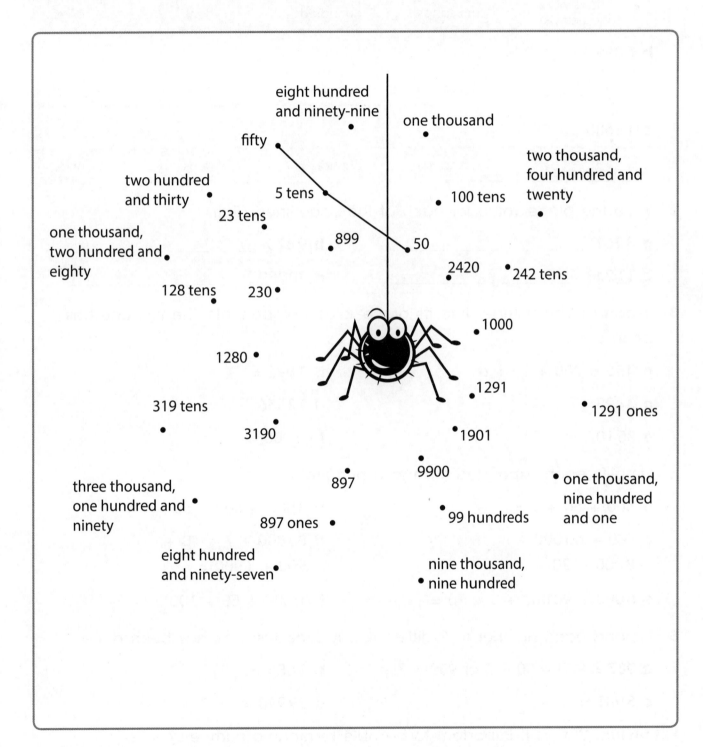

2 On another sheet of paper, create a spider web of your own.

Place Value (TRB pp. 28–31)
Number and place value Apply place value to partition, rearrange and regroup numbers to at least tens of thousands to assist calculations and solve problems.
(ACMNA073) AC

Expanding Numbers

1 Use the number expander to expand each of the numbers.

a 325

	tens of thousands		thousands		hundreds		tens		ones

b 6399

	tens of thousands		thousands		hundreds		tens		ones

c 12685

	tens of thousands		thousands		hundreds		tens		ones

2 Write the place value for each of the underlined digits.

a 3 2<u>5</u>1 _____

b <u>9</u>875 _____

c 11<u>1</u>248 _____

d <u>7</u>0789 _____

3 Expand each number into its place-value components. The first one has been done.

a 456 = 400 + 50 + 6

b 1092 =

c 3498 =

d 12456 =

e 35107 =

f 66000 =

4 Simplify each expanded form to a numeral.

a 500 + 60 + 1 =

b 3000 + 800 + 50 + 7 =

c 400 + 60000 + 8000 + 20 + 3 =

d 80000 + 7 + 40 + 900 + 1000 =

e 5000 + 90000 + 8 + 70 =

f 70000 + 50 + 100 =

5 Expand each number in 2 different ways. The first one has been done.

a 987 = 900 + 80 + 7 or 980 + 7

b 1056 =

c 5981 =

d 29990 =

Extension: Why is it important to be able to rename numbers?

Place Value (TRB pp. 28–31)
Number and place value Apply place value to partition, rearrange and regroup numbers to at least tens of thousands to assist calculations and solve problems
(ACMNA073) AC

Unit
3

STUDENT ASSESSMENT

1 Write the numbers in the correct columns on the place-value chart.

	tens of thousands	thousands	hundreds	tens	ones
a 4 798					
b 9 863					
c 10 256					
d 21 568					

2 Write the number that can be read on each number expander.

a

2	hundreds	3	6	ones

_____

b

5	2	thousands	4	2	1	ones

_____

3 Simplify each expanded form to a numeral.

a 30 000 + 6 000 + 400 + 20 + 1 =

b 80 000 + 1 + 70 =

4 Write the numbers in expanded form.

a 23 568 = _____

b 48 244 = _____

5 Rename Question 4a. 23 568 = _____

Extension: Write a 6-digit number. _____

Expand this number. _____

Rename this number. _____

On another sheet of paper, draw a representation of this number.

Unit
3

Place Value (TRB pp. 28–31)
Number and place value Apply place value to partition, rearrange and regroup numbers to at least tens of thousands to assist calculations and solve problems
(ACMNA073) **AC**

15

Lengths of Feet

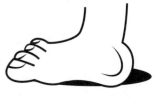

You will need: a ruler

Measure the length of the feet of 10 people in your classroom.
Collect the information in the table below.

Name	Length of foot in centimetres

1 Who had the **longest** foot in your classroom? _____

2 Who had the **shortest** foot in your classroom? _____

3 Were there any people with feet of the **same** length? _____

4 What is important to remember when measuring people's feet?

_____

Extension: Create a graph of the lengths
of people's feet in your classroom.

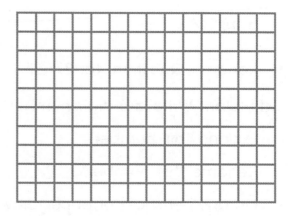

16 Unit **4** **Length and Temperature** (TRB pp. 32–35)
Using units of measurement Use scaled instruments to measure and
compare lengths, masses, capacities and temperatures
(ACMMG084) **AC**

Comparing Length

You will need: access to a range of measuring equipment such as rulers, measuring tapes and trundle wheels; a partner

Your task is to measure the lengths of 5 different lines on a play area such as a basketball court. Collect the information in the table below.

Description of line	Length in centimetres	Length in metres

Use the box to create a sketch of the play area you chose and the lines that you measured.

1 What was different between the answers in centimetres and those in metres? _____

_____

2 Which do you think would be more accurate? _____

_____

3 Did other groups who measured the same lines have the same or different answers? _____

_____

4 Which measuring tool did you prefer to use? Why? _____

_____

Unit 4

Length and Temperature (TRB pp. 32–35)
Using units of measurement Use scaled instruments to measure and compare lengths, masses, capacities and temperatures
(ACMMG084) **AC**

17

Temperature

1 Circle the thermometer with the **lowest** temperature.

a   b   c   d   e

2 Circle the thermometer with the **highest** temperature.

a   b   c   d   e

3 Colour each thermometer:

a   20°C   b   10°C   c   0°C   d   6°C   e   15°C

4 Write the temperature shown on each thermometer.

a   b   c   d   e

5 Circle the object that would be the hottest.

a   b   c   d

Unit **4**   **Length and Temperature** (TRB pp. 32–35)
Using units of measurement Use scaled instruments to measure and
compare lengths, masses, capacities and temperatures
(ACMMG084) **AC**

STUDENT ASSESSMENT

1 Measure each line and write the length in centimetres on the line.

a _____

b _____

c _____

d _____

2 Under each object, indicate whether its length would be measured in **centimetres** or **metres**.

a    **b**   **c**   **d**

_____   _____   _____

3 Rank each of these lengths from shortest (a) to longest (e).
Then record each length in metres.

Length (cm)	143 cm	120 cm	136 cm	150 cm	128 cm
Rank (a–e)					
Length (m)					

4 Write the temperature shown on each thermometer.

a    **b**    **c**

_____   _____   _____

5 Colour each thermometer:

a 25°C    **b** 8°C    **c** 17°C

Unit
4

Length and Temperature (TRB pp. 32–35)
Using units of measurement Use scaled instruments to measure and
compare lengths, masses, capacities and temperatures
(ACMMG084) AC

19

Kilograms or Grams

1 Look carefully at each illustration. Circle **grams** or **kilograms** to indicate the most suitable mass unit used to measure it.

a grams kilograms

b grams kilograms

c grams kilograms

d grams kilograms

e grams kilograms

f grams kilograms

2 How did you decide when to measure with grams, and when to measure with kilograms? _____

3 Order the objects from **least** mass (a) to the **greatest** mass (f).

1 kg   1.5 kg $\frac{3}{4}$ kg   2 kg   1.25 kg $\frac{1}{4}$ kg

_____   _____   _____   _____   _____   _____

4 Decide whether **grams** or **kilograms** should be used to weigh each thing.

a a man _____

b a mouse _____

c a wombat _____

d an ant _____

5 Read the scales to record each mass in kilograms.

_____   _____   _____   _____

Extension: How many kilograms are in each of the following?

a 3 000 grams _____   **b** 5 000 grams _____   **c** 8 000 grams _____

Unit **5**
Mass and Capacity (TRB pp. 36–39)
Using units of measurement Use scaled instruments to measure and compare lengths, masses, capacities and temperatures.
(ACMMG084) **AC**

Graphing Capacity

Students investigated the capacities of 12 different containers.
The graph shows their results.

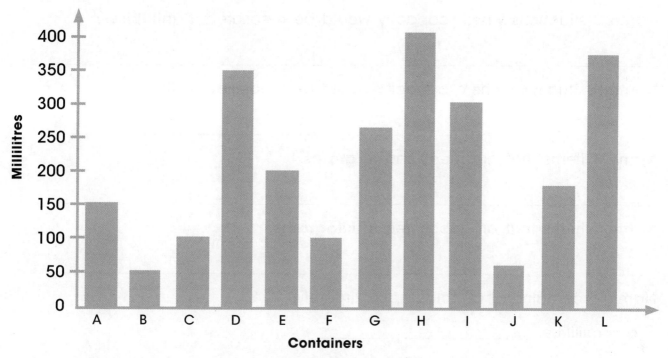

1 a Which container has the **greatest** capacity? _____

 b Which container has the **least** capacity? _____

 c Which 2 containers hold the same amount? _____

2 How many millilitres does:

 a container **A** hold? _____   b container **E** hold? _____

3 If you add containers **B**, **D**, **F** and **H**, what is the total capacity of all
 4 containers? _____

4 The combined capacity of containers **C** and **E** equals the capacity
 of which other container? _____

5 Rank the containers from **least** to **greatest** capacity.

_____

Extension: What is the total capacity of all of the containers in the graph?

_____

Unit
5
Mass and Capacity (TRB pp. 36–39)
Using units of measurement Use scaled instruments to measure and
compare lengths, masses, capacities and temperatures
(ACMMG084) **AC**

21

Mass and Capacity

Answer, or investigate and answer. (You may need access to a calculator.)

1 Name 3 situations where capacity would be measured in **millilitres**.

_____

2 Name 3 situations where capacity would be measured in **litres**.

_____

3 Name 3 items that are measured in **grams**.

_____

4 Name 3 items that are measured in **kilograms**.

_____

5 Name a container that holds approximately:

    a 5 millilitres _____

    b 1 litre _____

    c 500 millilitres _____

6 What does the **net weight** of an object mean?

_____

7 Imagine that you have to fill a swimming pool. You have no hose, and you have to carry the water to the pool! How might you fill the pool?

_____

Extension: The dimensions for a standard in-ground pool are: 10 metres long, 4 metres wide and 1.5 metres deep.

    a What is the **capacity** (or volume) of this pool?

_____

    b If 1 litre of water weighs 1 kilogram, then what would be the **mass** of the water if the pool was filled?

_____

22   Unit 5   **Mass and Capacity** (TRB pp. 36–39)
Using units of measurement Use scaled instruments to measure and compare lengths, masses, capacities and temperatures
(ACMMG084)

STUDENT ASSESSMENT

1 Order the items from **lightest** to **heaviest**.

a 800 grams **b** 1200 grams

c 50 grams **d** 40 grams

_____

2 Under each object, write whether the mass would be measured in **grams** or **kilograms**.

a **b** **c** **d**

_____ _____ _____ _____

3 Explain what **capacity** means. _____

_____

4 Record the amount in each measuring jug.

a **b** **c**

_____ _____ _____

5 How many of this container would be needed to fill each of the following containers?

a **b** **c**

_____ _____ _____

Unit
5

Mass and Capacity (TRB pp. 36–39)
Using units of measurement Use scaled instruments to measure and
compare lengths, masses, capacities and temperatures.
(ACMMG084) **AC**

23

Number Sequences of 3

1 Complete the number sequences.

a 3, 6, 9, 12, ____, ____, ____, ____   **b** 10, 13, 16, 19, ____, ____, ____, ____

c 35, 38, 41, 44, ____, ____, ____, ____   **d** 100, 97, 94, 91, ____, ____, ____, ____

2 Fill in the missing numbers.

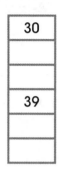

a
6
12

b

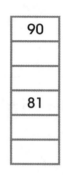

b
30
39

c

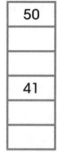

c
90
81

d

d
50
41

3 Write the first 5 terms in each number sequence.

a Start at 20 and count forwards by 3s.

_____, _____, _____, _____, _____

b Start at 105 and count forwards by 3s.

_____, _____, _____, _____, _____

c Start at 99 and count backwards by 3s.

_____, _____, _____, _____, _____

4 Show the number sequences on the number lines.

a Start at 15 and count forwards by 3s.

⟵──────────────────────────────────⟶

b Start at 60 and count forwards by 3s.

⟵──────────────────────────────────⟶

5 Draw a picture to show counting by 3s.

24 Unit **6** **Number Sequences: 3s, 6s and 9s** (TRB pp. 40–43)
Number and place value Investigate number sequences involving multiples of 3, 4, 6, 7, 8, and 9
(ACMNA074) **AC**

AFL Football Scores

1 Show the number sequence when starting at 0 and counting forwards by 6s.

_____, _____, _____, _____, _____, _____, _____, _____, _____

2 Use the number sequence in Question 1 to help you complete the AFL football scores for Round 16. You have to:

a Find the total score for each team.
(Remember it is the number of goals × 6 + the number of points.)

b Colour all of the winning teams and scores in appropriate colours.

West Coast Eagles vs Geelong					
14 goals	12 points	(14 × 6) + 12 = 96	13 goals	10 points	
Hawthorn vs Brisbane Lions					
16 goals	9 points		9 goals	9 points	
Fremantle vs Melbourne					
4 goals	12 points		15 goals	16 points	
Essendon vs Richmond					
15 goals	15 points		9 goals	12 points	
Collingwood vs North Melbourne					
22 goals	15 points		3 goals	12 points	
Port Adelaide vs St Kilda					
8 goals	5 points		17 goals	7 points	
Western Bulldogs vs Carlton					
14 goals	12 points		9 goals	15 points	

Unit 6 Number Sequences: 3s, 6s and 9s (TRB pp. 40–43)
Number and place value Investigate number sequences involving multiples of 3, 4, 6, 7, 8, and 9
(ACMNA074) **AC**

25

Number Sequences of 9

1 Complete the number sequences counting by 9s.

a 9, 18, 27, 36, _____, _____, _____, _____, _____

b 45, 54, 63, 72, _____, _____, _____, _____, _____

c 108, 99, 90, 81, _____, _____, _____, _____, _____

d 72, 63, 54, 45, _____, _____, _____, _____

2 Try some trickier number sequences.

a 10, 19, 28, 37, _____, _____, _____, _____, _____

b 150, 159, 168, 177, _____, _____, _____, _____, _____

c 200, 191, 182, 173, _____, _____, _____, _____

d 125, 116, 107, 98, _____, _____, _____, _____, _____

3 What patterns do you notice?

_____

4 Try completing the number sequences with adding 10 and subtracting 1.
For example: 9 + 10 = 19 – 1 = 18, and so on.

a 20, 29, 38, 47, _____, _____, _____, _____, _____

b 85, 94, 103, 112, _____, _____, _____, _____, _____

c 110, 119, 128, 137, _____, _____, _____, _____, _____

5 Complete the spiral number sequences.

a  b  c

6 Explain how you worked out the number sequences from Question 5.

_____

26 | Unit 6 | **Number Sequences: 3s, 6s and 9s** (TRB pp. 40–43)
Number and place value Investigate number sequences involving multiples of 3, 4, 6, 7, 8, and 9
(ACMNA074) **AC**

STUDENT ASSESSMENT

1 On the 100 chart, colour:

a number sequence by 3s in green.

b number sequence by 6s in red.

c number sequence by 9s in orange.

Describe some of the patterns.

_____

_____

1	2	3	4	5	6	7	8	9	10
11	12	13	14	15	16	17	18	19	20
21	22	23	24	25	26	27	28	29	30
31	32	33	34	35	36	37	38	39	40
41	42	43	44	45	46	47	48	49	50
51	52	53	54	55	56	57	58	59	60
61	62	63	64	65	66	67	68	69	70
71	72	73	74	75	76	77	78	79	80
81	82	83	84	85	86	87	88	89	90
91	92	93	94	95	96	97	98	99	100

2 Show the following number sequences on the number lines, starting at the number 2.

a 3s

b 6s

3 Explain the link between the 3s number sequence and the 6s number sequence. _____

_____

4 Complete the number sequences.

a 9, 12, 15, 18, _____, _____, _____, _____, _____

b 90, 99, 108, 117, _____, _____, _____, _____, _____

c 40, 46, 52, 58, _____, _____, _____, _____, _____

Extension: Complete the number sequence.

100, 112, 124, 136, _____, _____, _____, _____, _____

Explain how you worked it out.

_____

_____

Unit 6
Number Sequences: 3s, 6s and 9s (TRB pp. 40–43)
Number and place value Investigate number sequences involving multiples of 3, 4, 6, 7, 8, and 9
(ACMNA074) **AC**

27

Number Sequences of 4

DATE:

1 Complete the number sequences.

a 4, 8, 12, 16, _____, _____, _____, _____, _____

b 40, 44, 48, 52, _____, _____, _____, _____, _____

c 35, 39, 43, 47, _____, _____, _____, _____, _____

d 100, 96, 92, 88, _____, _____, _____, _____, _____

2 Use words to describe each number sequence.

a 10, 14, 18, 22, _____

b 120, 124, 128, 132, _____

c 83, 79, 75, 71, _____

d 50, 46, 42, 38, _____

3 Write the first 5 terms in each number sequence.

a Start at 60 and count forwards by 4s. _____, _____, _____, _____, _____

b Start at 105 and count forwards by 4s. _____, _____, _____, _____, _____

c Start at 31 and count forwards by 4s. _____, _____, _____, _____, _____

d Start at 97 and count backwards by 4s. _____, _____, _____, _____, _____

4 Show the 2 number sequences on the same number line.

a Start at 15 and count forwards by 2s.

b Start at 15 and count forwards by 4s.

15

c What do you notice? _____

5 Explain how using doubles helps you to remember the 4s number sequence.

_____

Extension: If we doubled the 4s number sequence, what would happen?

_____

Number Sequences: 4s, 8s and 7s (TRB pp. 44–47)
Number and place value Investigate number sequences involving multiples of 3, 4, 6, 7, 8, and 9
(ACMNA074) **AC**

Number Sequence Patterns

You will need: a ruler, a calculator (for Extension)

On each circle, draw a line from one number to the next in the number sequence. When you reach numbers higher than 9, use the last digit in each number.

For example: for the 4s number sequence 4, 8, 12, 16... you would draw lines from 4 to 8 to 2 to 6, and so on.

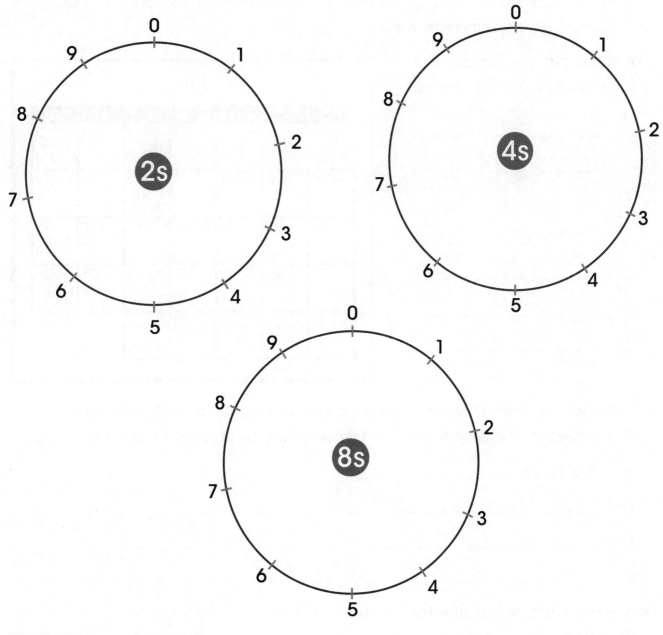

What did you discover? _____

Extension: On a sheet of paper, investigate number sequences of 12 and 16.

Unit 7

Number Sequences: 4s, 8s and 7s (TRB pp. 44–47)
Number and place value: Investigate number sequences involving multiples of 3, 4, 6, 7, 8, and 9
(ACMNA074) AC

29

Number Sequences of 7

DATE:

1 Look at the page from a calendar.

a Starting at 7, shade each of the days in the 7s number sequence in blue.

b Starting at 4, shade each of the days in the 7s number sequence in orange.

c Starting at 1, shade each of the days in the 7s number sequence in green.

d Select a different starting point and shade each of the days in the 7s number sequence in red.

e What did you notice in the colouring patterns?

_____

_____

_____

_____

_____

_____

July

MON	TUES	WED	THURS	FRI	SAT	SUN
		1	2	3	4	5
6	7	8	9	10	11	12
13	14	15	16	17	18	19
20	21	22	23	24	25	26
27	28	29	30	31		

2 List each number sequence from above and continue with the next 5 numbers in the sequence. The first one has been started for you.

a 7, 14, 21, 28, _____, _____, _____, _____, _____

b _____, _____, _____, _____, _____, _____, _____, _____, _____

c _____, _____, _____, _____, _____, _____, _____, _____, _____

d _____, _____, _____, _____, _____, _____, _____, _____, _____

Extension: Write any strategies you used to help you work out the 7s number sequence.

_____

30

Unit **7**

Number Sequences: 4s, 8s and 7s (TRB pp. 44–47)
Number and place value Investigate number sequences involving multiples of 3, 4, 6, 7, 8, and 9
(ACMNA074) **AC**

DATE:

STUDENT ASSESSMENT

1 On the number line, show the number sequences:

a by 4s in green. **b** by 8s in red. **c** by 7s in blue.

0

2 Complete the number sequences.

a Start at 1 and count by 7s. _____, _____, _____, _____, _____

b Start at 4 and count by 8s. _____, _____, _____, _____, _____

c Start at 3 and count by 4s. _____, _____, _____, _____, _____

3 Explain the link between the 4s number sequence and the 8s number sequence. _____

4 Describe each number sequence in words.

a 2, 6, 10, 14, 18, 22

_____

b 10, 18, 26, 34, 42

_____

c 11, 18, 25, 32, 39

_____

5 Complete the number sequence.

100, 107, 114, 121, 128, _____, _____, _____, _____

Extension: Write a number sequence starting at 100 and counting back by 8s. _____

Explain how you worked it out. _____

_____

Unit
7
Number Sequences: 4s, 8s and 7s (TRB pp. 44–47)
Number and place value Investigate number sequences involving multiples of 3, 4, 6, 7, 8, and 9
(ACMNA074) **AC**

31

Regular Shapes

1 Write a description for each regular shape.

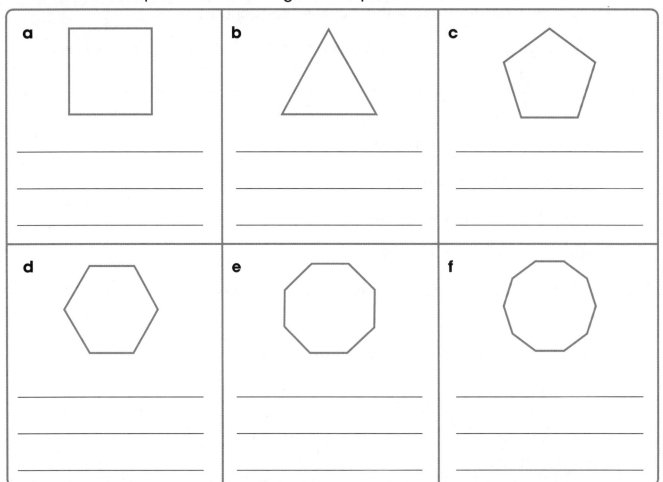

2 Circle the shape in each set that is the odd one out (not a regular shape).

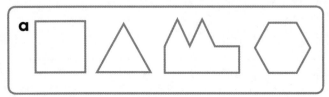

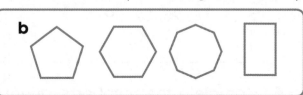

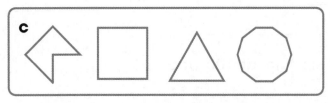

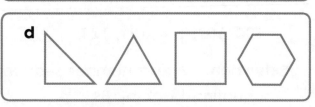

Extension: Can you name these shapes?

Regular or Irregular?

1 Colour all the shapes that are regular.

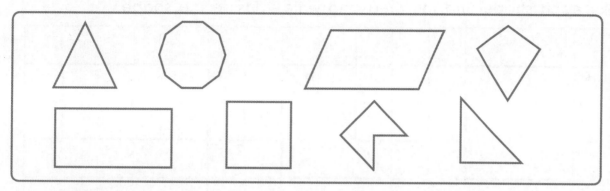

2 Circle the shape in each set that is the odd one out
(not a regular shape).

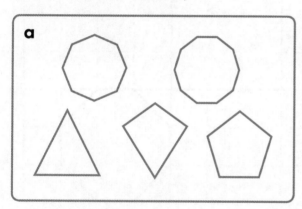

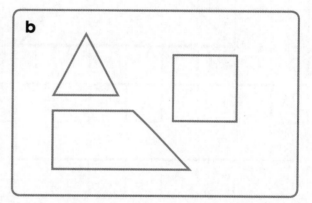

3 Using the grid paper as a guide, draw 3 different irregular shapes and
2 regular shapes. Label each shape **regular** or **irregular**.

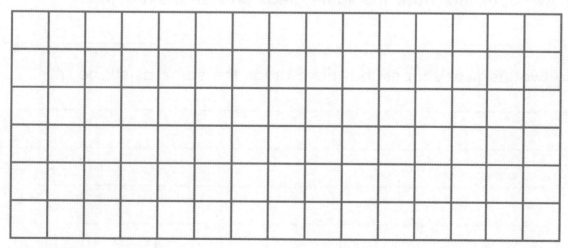

4 Find the area of each shape by counting squares. Write the answer on
your shape in cm².

Unit **8**

Regular Shapes (TRB pp. 48–51)
Shape Compare the areas of regular and irregular shapes by informal means
(ACMMG087) **AC**

33

Comparing Regular and Irregular

1 For each shape, find the area and write it inside the shape.

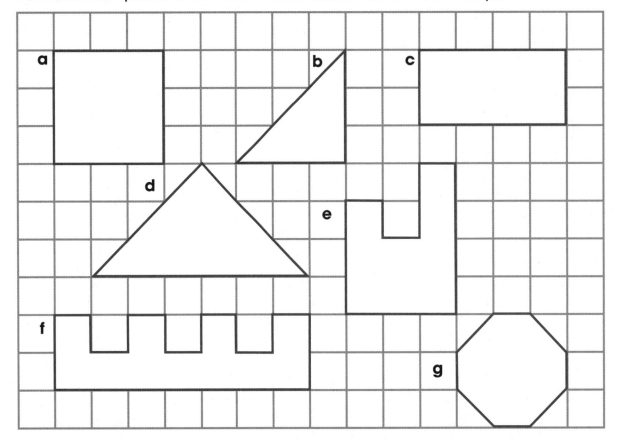

2 Shade each **regular** shape red and each **irregular** shape green.

3 Find the shapes that have the same areas and list them below.

_____

4 Create two irregular shapes that have areas the same as shape **g**.

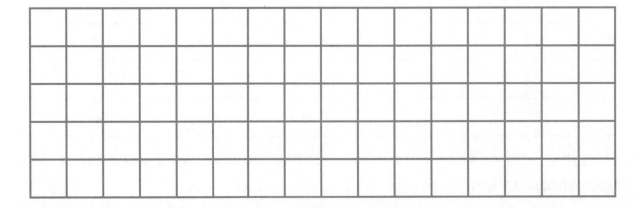

Unit 8 **Regular Shapes** (TRB pp. 48–51)
Shape Compare the areas of regular and irregular shapes by informal means
(ACMMG087) **AC**

STUDENT ASSESSMENT

DATE:

1 Label each shape **regular** or **irregular**.

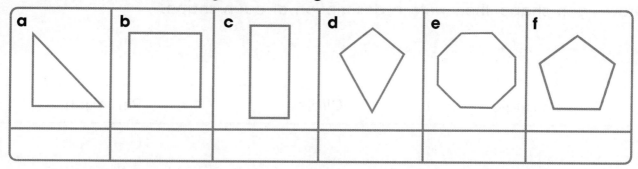

a	b	c	d	e	f

2 Describe what features make a regular shape, e.g. a pentagon.

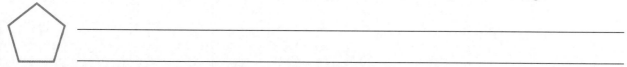

3 Find and record the area of each shape.

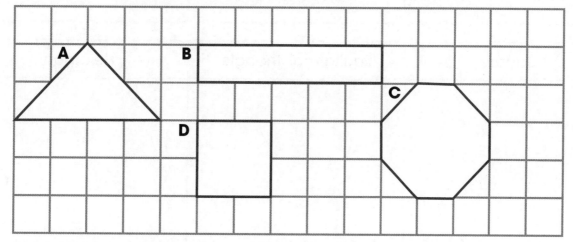

a Which two shapes have the **same** area? _____

b Which shape has the **largest** area? _____

c What is the area of the **smallest** shape? _____

Extension: Draw an irregular shape that is 4 cm².

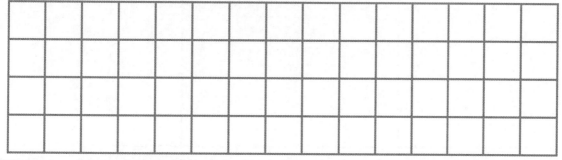

Unit
8
Regular Shapes (TRB pp. 48–51)
Shape Compare the areas of regular and irregular shapes by informal means
(ACMMG087) AC

35

2D Shapes

Look at the names of 2D shapes.

Draw each shape, then write a description of each shape.

Rectangle	Circle	Pentagon

Square	Equilateral triangle	Hexagon

Trapezium	Parallelogram	Semi-circle

Features of Tangrams

Here is a tangram.

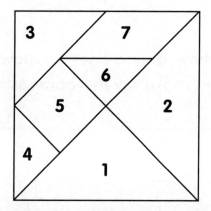

1   Complete the table with the features of each numbered shape in the tangram.

Number	Name of shape	Number of edges	Number of vertices	Description of the shape
1				
2				
3				
4				
5				
6				
7				

2   Here is a different tangram.

Describe how it is different from the tangram above.

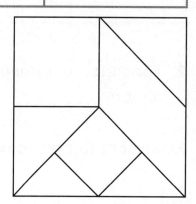

_____

_____

_____

Unit 9   **2D Shapes** (TRB pp. 52–55)
Shape Compare and describe two dimensional shapes that result from combining and splitting common shapes, with and without the use of digital technologies
(ACMMG088)   **AC**

37

Pulling Composite Shapes Apart

1 For each composite shape, draw dotted lines to show the common shapes that could have been used to make up the composite shape.

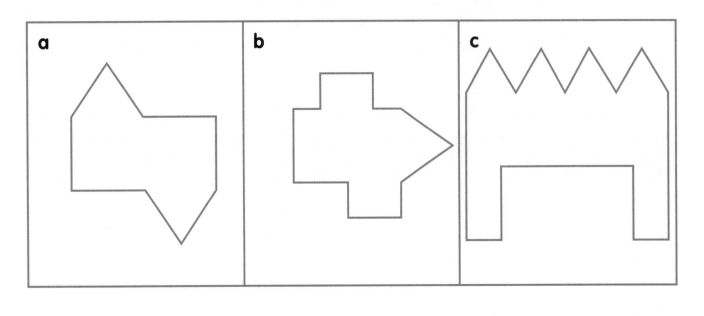

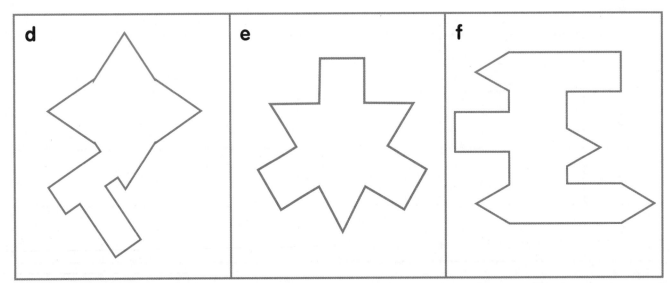

2 For shape **d**, list the names of all the shapes that make up the composite shape. _____

Extension: Design some composite shapes for your friends to pull apart!

DATE:

STUDENT ASSESSMENT

1 Complete the table by drawing the shapes and adding in the details.

Draw 2D shape				
Name	square	rectangle	triangle	hexagon
Number of edges				
Number of corners				

2 Using the grid paper, draw a composite shape made up of 2 rectangles and 1 triangle.

3 By drawing in dotted lines, show what common 2D shapes make up the composite shapes.

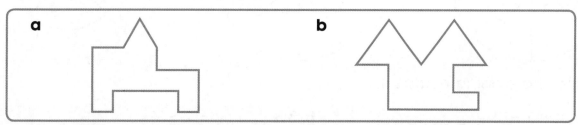

a

b

Extension: Draw a composite shape that has an area of 10 cm².

Unit 9

2D Shapes (TRB pp. 52–55)

Shape Compare and describe two dimensional shapes that result from combining and splitting common shapes, with and without the use of digital technologies

(ACMMG088) **AC**

39

Times Tables

1 Complete the equations.

a 4 × 5 =   **b** 3 × 8 =   **c** 7 × 3 =

d 10 × 4 =   **e** 5 × 5 =   **f** 6 × 9 =

g 7 × 1 =   **h** 2 × 10 =   **i** 8 × 7 =

j 9 × 4 =   **k** 3 × 6 =   **l** 4 × 4 =

m 5 × 9 =   **n** 6 × 8 =   **o** 1 × 1 =

2 Complete the times tables grids.

a

×	2	5	7	9	10
3					

b

×	4	1	9	8	3
6					

c

×	4	2	8	6	9
9					

d

×	10	8	7	1	2
5					

3 Complete the wheels.

a 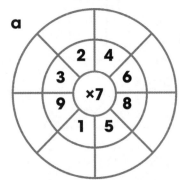   **b** 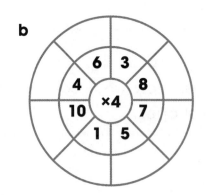   **c**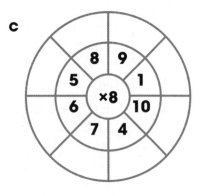

4 Find the missing numbers.

a 4 × 5 = 2 × _____   **b** 3 × _____ = 2 × 6

c 5 × 8 = _____ × 10   **d** 2 × _____ = 3 × 6

e 8 × _____ = 4 × 2   **f** _____ × 4 = 6 × 6

Extension: Explain the link between the 2, 4 and 8 times tables.

_____

_____

9 × Tables

1 Try this trick for the 9 × tables. Using your fingers:

Hold out both your hands in front of you with your fingers spread out. Let's look at 9 × 4. Bend down your 4th finger from the left (9 × 3 would be the 3rd finger, and so on). The number of fingers in front (to the left) of the bent finger is the first digit in the answer: 3. The number of fingers after (to the right of) the bent finger is the second digit: 6. Therefore 9 × 4 = 36.

Note: this technique works best up to 10 × 9.
Try completing the tables with the finger trick.

a 4 × 9 =   **b** 3 × 9 =   **c** 7 × 9 =

2 Here is another trick:

The tens digit in the answer is always 1 less than the multiplying number. The two digits of the answer always add to 9.

So, for 3 × 9, 3 − 1 = 2 and 2 + 7 = 9, so the answer is 27.
Try these:

a 5 × 9 =   **b** 2 × 9 =   **c** 8 × 9 =   **d** 9 × 6 =

3 Here is a final trick:

10 × take away the number. For example: 4 × 9

Multiply the number by 10: 4 × 10 = 40

Then take away the number from the answer: 40 − 4 = 36

So, 4 × 9 = 36.

Try these:

a 9 × 9 =   **b** 7 × 9 =   **c** 1 × 9 =

d 9 × 4 =   **e** 9 × 6 =   **f** 9 × 10 =

4 Which trick do you prefer? _____

Unit 10 Multiplication Facts (Times Tables) (TRB pp. 56–59)
Number and place value Recall multiplication facts up to 10 × 10 and related division facts
(ACMNA075) **AC**

41

More Tables

1 Represent each table on a number line.

a 4 × 7 =

<-->

b 7 × 6 =

<-->

c 2 × 5 =

<-->

2 Colour the answers for the tables on the 100 chart.

a Blue: 4 ×

b Red: 8 ×

c Green: 6 ×

What patterns do you notice?

_____

1	2	3	4	5	6	7	8	9	10
11	12	13	14	15	16	17	18	19	20
21	22	23	24	25	26	27	28	29	30
31	32	33	34	35	36	37	38	39	40
41	42	43	44	45	46	47	48	49	50
51	52	53	54	55	56	57	58	59	60
61	62	63	64	65	66	67	68	69	70
71	72	73	74	75	76	77	78	79	80
81	82	83	84	85	86	87	88	89	90
91	92	93	94	95	96	97	98	99	100

3 True or false (T or F)?

a 7 × 7 = 8 × 7 _____   **b** 7 × 4 = 3 × 8 _____   **c** 5 × 9 = 9 × 5 _____
d 7 × 4 = 3 × 6 _____   **e** 2 × 6 = 4 × 3 _____   **f** 5 × 8 = 10 × 4 _____

4 Find the missing numbers.

a _____ × 10 = 80   **b** 4 × _____ = 36   **c** 8 × _____ = 24

d _____ × 6 = 54   **e** 7 × _____ = 35   **f** _____ × 5 = 20

Extension: Find the total cost of:

a 6 movie tickets at $8 each. _____

b 7 chocolate bars at $3 each. _____

c 5 sandwiches at $4 each. _____

d 9 books at $10 each. _____

Unit
10

STUDENT ASSESSMENT

1 How would you rate your recall of multiplication facts (tables)?

poor   **OK**   **good**   **very good**

2 List the tables you need more practice with.

_____

3 Complete the tables.

a 4 × 3 =   **b** 7 × 2 =   **c** 10 × 10 =   **d** 5 × 6 =

e 8 × 6 =   **f** 9 × 4 =   **g** 3 × 8 =   **h** 6 × 6 =

4 Fill in the missing numbers.

a _____ × 6 = 12   **b** 9 × _____ = 45   **c** 10 × _____ = 60

d _____ × 8 = 16   **e** 18 = _____ × 9   **f** 64 = _____ × 8

5 Explain how you found the answer to Question 4b.

_____

6 Draw 3 arrays that give the answer 24.

Extension: How many legs are on 8 rabbits? _____

Unit
10

Multiplication Facts (Times Tables) (TRB pp. 56–59)
Number and place value Recall multiplication facts up to 10 × 10 and related division facts
(ACMNA075) **AC**

43

Using Your Multiplication Grid

You will need: BLM 3 'Tables Chart 1', BLM 4 'Tables Chart 2'

1 Complete the tables using your multiplication grid.

a 2 × 5 =   **b** 3 × 6 =   **c** 8 × 3 =   **d** 9 × 4 =   **e** 6 × 6 =

f 7 × 9 =   **g** 9 × 1 =   **h** 5 × 10 =   **i** 4 × 7 =   **j** 9 × 3 =

k 8 × 6 =   **l** 2 × 2 =   **m** 5 × 7 =   **n** 8 × 8 =   **o** 1 × 1 =

2 What did you notice as you completed the questions with your grid?

_____

3 Look at the tables 1 × 1; 2 × 2; 3 × 3; 4 × 4 and so on. What happens on the multiplication grid?

_____

4 **a** What happens when you look at complementary tables such as 3 × 2 and 2 × 3? _____

b Give 3 more examples of this. _____

5 Write an equation and use the multiplication grid to help solve:

a the number of legs on 9 pandas.   **b** the number of legs on 5 spiders.

_____   _____

c the number of legs on 10 insects.   **d** the number of legs on 7 chickens.

_____   _____

6 Use the multiplication grid to help solve the wheels.

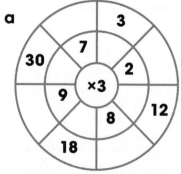

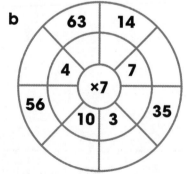

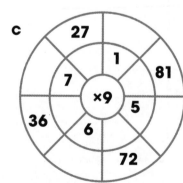

Extension: List all the patterns you can see in the multiplication grid.

44   Unit 11   **Multiplication Facts and Related Division Facts** (TRB pp. 60–63)
Number and place value Recall multiplication facts up to 10 × 10 and related division facts
(ACMNA075) AC

Fact Families

1 Fill in the missing number in each fact family.

a $4 \times 3 =$ _____

b $4 \times$ _____ $= 12$

c _____ $\times 3 = 12$

d $3 \times 4 =$ _____

e $3 \times$ _____ $= 12$

f _____ $\times 4 = 12$

g $12 \div 4 =$ _____

h $12 \div$ _____ $= 3$

i _____ $\div 4 = 3$

j $12 \div 3 =$ _____

k $12 \div$ _____ $= 4$

l _____ $\div 3 = 4$

m $12 = 4 \times$ _____

n $12 =$ _____ $\times 3$

o _____ $= 4 \times 3$

p $12 = 3 \times$ _____

q $12 =$ _____ $\times 4$

r _____ $= 3 \times 4$

s $3 = 12 \div$ _____

t $3 =$ _____ $\div 4$

u _____ $= 12 \div 4$

v $4 = 12 \div$ _____

w $4 =$ _____ $\div 3$

x _____ $= 12 \div 3$

> Sometimes an array can be helpful!
>
> ● ● ● ●
> ● ● ● ●
> ● ● ● ●

2 Are these equations part of the family? Why or why not?

$3 \div 12 = 4$ $\qquad$ $4 \div 12 = 3$ $\qquad$ $4 \div 3 =$ _____ $\qquad$ _____ $\times 3 = 12$

_____

3 Complete a fact family for the equation $7 \times 3 = 21$. How many related equations can you find?

_____

4 Write a multiplication equation that equals 50. Find all of the related division equations of the fact family.

_____

Extension: Try investigating the fact family of a number such as 1 100.

Unit 11 **Multiplication Facts and Related Division Facts** (TRB pp. 60–63)
Number and place value Recall multiplication facts up to 10 × 10 and related division facts
(ACMNA075) **AC**

45

Division with Arrays

1 Draw arrays for each of the multiplication equations.

2 Use the arrays to help find the related division equations.

Multiplication equation	Array	Division equation
a 6 × 4		
b 4 × 5		
c 8 × 2		
d 9 × 4		
e 2 × 4		
f 6 × 2		

Unit 11 **Multiplication Facts and Related Division Facts** (TRB pp. 60–63)
Number and place value Recall multiplication facts up to 10 × 10 and related division facts
(ACMNA075) AC

DATE:

STUDENT ASSESSMENT

1 Draw an array for 3 × 10.

2 List all the multiplication equations related to the array.

_____

3 List all the division equations related to the array.

_____

4 Look at the picture.

a Give 3 multiplication equations related to the array.

_____

b Now give 3 division equations related to the array.

_____

5 Look at the equation 24 ÷ 6.

a Show the equation 24 ÷ 6 on the grid paper.

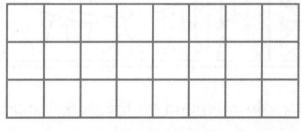

b Now show the equation 24 ÷ 4 on the grid paper.

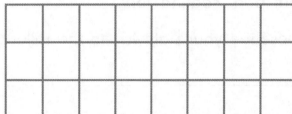

c What do you notice?

Extension: On another piece of paper, list all the related equations you can think of for the fact family of 12 × 12 = 144.

Unit
11

Multiplication Facts and Related Division Facts (TRB pp. 60–63)
Number and place value Recall multiplication facts up to 10 × 10 and related division facts
(ACMNA075) **AC**

47

At the Fun Park

Look at the map of a fun park.

Scale: 1cm = 10m

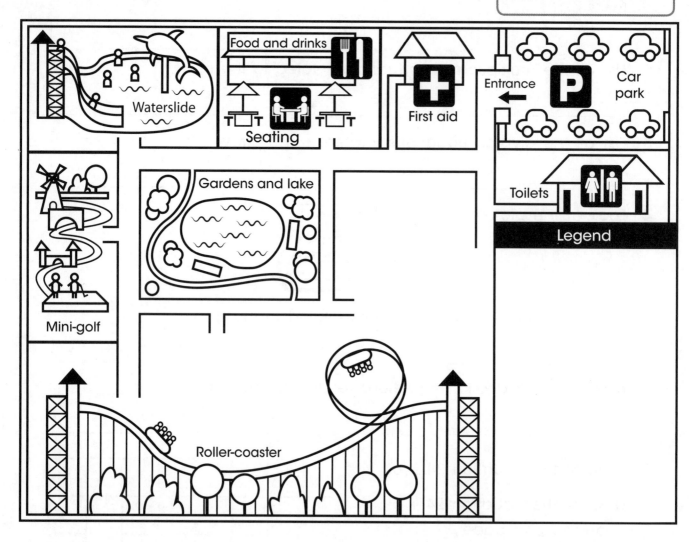

1   Create a legend for the park. Include facilities such as toilets, first aid, car park, seating, food.

2   Describe your pathway to the roller-coaster from the car park.

_____

_____

Extension: Add some more features to the fun park and include them in your legend.

The Science Centre

Look at the map of the Science Centre.

Scale: 1cm = 10m

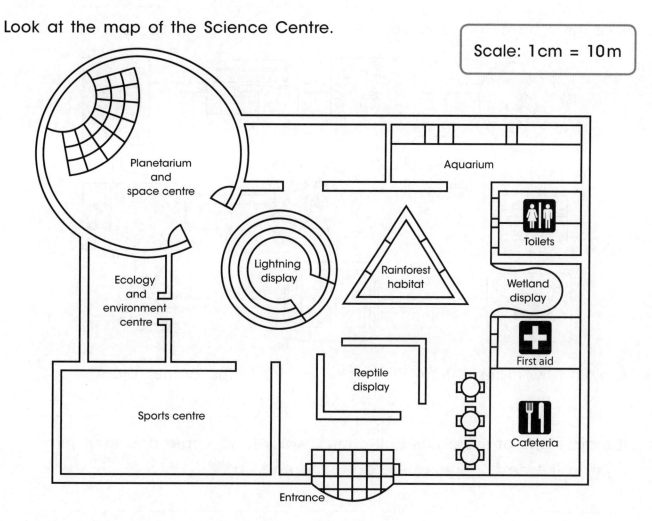

1 What is the scale of the map? _____

2 How far are the toilets from the lightning display? _____

3 How far is the planetarium from the entrance? _____

4 How far is the cafeteria from the sports centre? _____

5 Create a path through the Science Centre on the map. How long is your path? _____

6 Describe how you worked out the length of the path. _____

Extension: Add a new display to the map in the empty space. Describe how far the new display is from the entrance and the cafeteria.

_____

Unit
12
Mapping (TRB pp. 64–67)
Location and transformation Use simple scales, legends and directions to interpret information contained in basic maps
(ACMMG090) **AC**

49

Things to Do

Look at the map of things to do.

Scale: 1 cm = 30 m

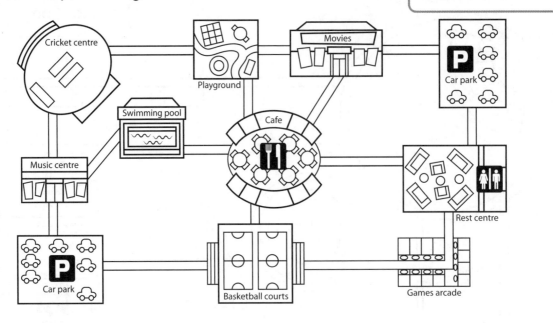

1 Explain how to travel from the cricket centre to the games arcade.

_____

2 If India starts at the basketball courts, walks to the cafe and then turns
   right to the next venue, where does she end up?

_____

3 If Kai is to meet his mum at the car park closest to the music centre, and
   he is at the movies, explain how he will get there.

_____

4 Alex is at the games arcade when he receives a call to meet Tim at the
   playground. Explain which path Alex might follow.

_____

5 Is there another way Alex could go? Explain.

_____

Extension: Colour three things you would like to do on the map. On another
sheet of paper, explain how you would get to these things if you
were at the car park closest to the movies.

50 Unit 12 **Mapping** (TRB pp. 64–67)
Location and transformation Use simple scales, legends and directions to interpret
information contained in basic maps
(ACMMG090) AC

STUDENT ASSESSMENT

Look at the map of a zoo.

Scale: 1 cm = 25 m

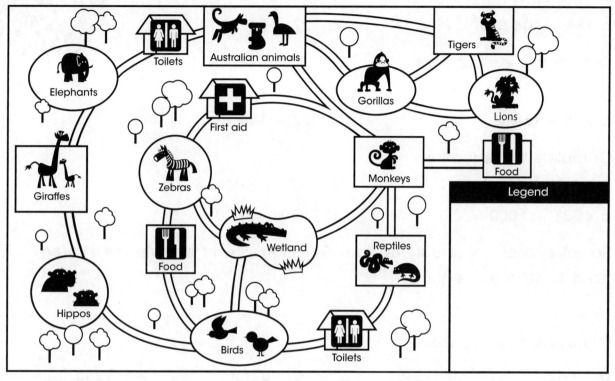

1 Add:

a a tree next to the lions' enclosure on the right.

b an ice-cream stand next to the reptiles on the left.

c an information centre in the middle of the zoo.

2 Create a legend for the zoo.

3 Explain how you would walk from the lions to the birds.

_____

_____

4 Is there more than one way to get from the lions to the birds? Give another example.

_____

_____

5 What is the scale of the map? _____

Unit **12** **Mapping** (TRB pp. 64–67)
Location and transformation Use simple scales, legends and directions to interpret
information contained in basic maps
(ACMMG090) **AC**

51

Addition

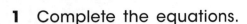

1 Complete the equations.

a 58 + 22 =   **b** 74 + 26 =

c 156 + 104 =   **d** 227 + 33 =

2 Complete the equations.

a 52 + 39 =   **b** 142 + 47 =

c 216 + 169 =   **d** 237 + 144 =

3 Complete the equations.

a 5 260 + 2 000 =   **b** 1 357 + 8 000 =

c 4 531 + 1 000 =   **d** 7 849 + 2 000 =

4 Select one of the equations from Question 2, and explain the strategy you used to solve the equation.

_____

5 Complete the equations.

a 432	**b** 249	**c** 8 862	**d** 3 205
+ 156	+ 530	+ 1 037	+ 4 691

6 Fill in the missing digits to complete the equations.

a   2 1 ☐
+ 4 ☐ 6
☐ 3 8

b   7 ☐ 4
+ ☐ 3 ☐
9 7 5

c   2 ☐ 6 ☐
+ ☐ 1 0 3
5 8 ☐ 7

d   6 ☐ 4 ☐
+ 3 5 ☐ 3
☐ 7 8 4

7 Solve the addition problems.

a Find the total of 8 543 and 1 352. _____

b Find 3 648 plus 1 221. _____

c Solve the addition of 6 395 and 2 403. _____

d Find the sum of 4 236 and 1 358. _____

Extension: Consecutive numbers follow one after the other. Find the sum of the consecutive numbers 3 267, 3 268 and 3 269. _____

52 **Unit 13** Addition and Subtraction (TRB pp. 68–71)
Number and place value Apply place value to partition, rearrange and regroup numbers to at least tens of thousands to assist calculations and solve problems
(ACMNA073) **AC**

Subtraction

1 Complete the equations.

a 58 – 23 =   **b** 67 – 43 =   **c** 156 – 122 =

2 Complete the grids.

a

–	56	75	68	95
21				

b

–	58	99	138	147
37				

c

–	257	368	497	559
53				

d

–	185	199	388	574
74				

3 Complete the equations.

a 46
    – 27
    ‾‾‾‾

b 86
    – 69
    ‾‾‾‾

c 856
    – 65
    ‾‾‾‾

d 695
    – 67
    ‾‾‾‾

4 Try these more challenging subtraction problems.

a Find 945 minus 37. _____

b Find the difference between 276 and 185. _____

c Find 46 less than 409. _____

d How much greater than 56 is 149? _____

5 Fill in the missing digits to complete the equations.

a 1 5 6 9
   – ☐ 3 ☐
   ‾‾‾‾‾‾‾
     1 1 ☐ 4

b 7 3 2 4
   – ☐ 2 ☐
   ‾‾‾‾‾‾‾
     7 1 ☐ 3

c 7 5 2 9
   –2 0 ☐ ☐
   ‾‾‾‾‾‾‾
     ☐ 5 1 0

6 Find the difference between:

a $181 and $95 _____   **b** $655 and $487 _____

c $1056 and $875 _____   **d** $5465 and $3282 _____

Extension: On another sheet of paper, write three different subtraction
  equations using 1256, 1386 and 2438. Solve them.

Unit **13** Addition and Subtraction (TRB pp. 68–71)
Number and place value Apply place value to partition, rearrange and regroup numbers to at least
tens of thousands to assist calculations and solve problems
(ACMNA073) **AC**

53

Addition and Subtraction Number Lines

1 Use the open number lines to solve the equations. Start with the largest number.

a 234 + 75 =

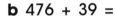

b 476 + 39 =

c 167 + 242 =

d 235 + 288 =

2 Use the open number lines to solve the equations.

a 256 – 43 =

b 297 – 38 =

c 391 – 167 =

3 Use the open number lines to find the difference between:

a 468 and 537

b 821 and 672

Extension: Use an open number line to show 125 + 36 + 72 =

Unit 13 Addition and Subtraction (TRB pp. 68–71)
Number and place value Apply place value to partition, rearrange and regroup numbers to at least tens of thousands to assist calculations and solve problems
(ACMNA073) **AC**

Unit

STUDENT ASSESSMENT

DATE:

1 Complete the addition equations.

   a 352 + 28 =   **b** 456 + 93 =   **c** 456 + 72 =

   d 243   **e** 349   **f** 458

   + 128   + 147   + 171

2 Complete the subtraction equations.

   a 114 – 39 =   **b** 248 – 63 =   **c** 597 – 128 =

   d 219   **e** 473   **f** 597

   – 116   – 152   – 189

3 Describe how you solved Question 2c. What strategy or strategies did you use?

_____

4 Complete the addition and subtraction patterns.

   a 2 + 3 =   **b** 8 – 2 =   **c** 4 + 7 =

      20 + 30 =   80 – 20 =   40 + 70 =

      200 + 300 =   800 – 200 =   400 + 700 =

      2 000 + 3 000 =   8 000 – 2 000 =   4 000 + 7 000 =

5 Use the open number lines to solve the equations.

   a 214 + 49 =

   b 325 – 52 =

Addition and Subtraction (TRB pp. 68–71)
Number and place value Apply place value to partition, rearrange and regroup numbers to at least tens of thousands to assist calculations and solve problems
(ACMNA073)

55

Perimeter of Irregular Shapes

You will need: a ruler, BLM 26 'Square Dot Paper'

1 Find the perimeter of each shape.

a

b

c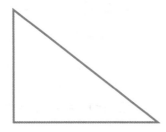

2 Find the perimeter of each irregular shape.

a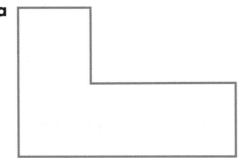

b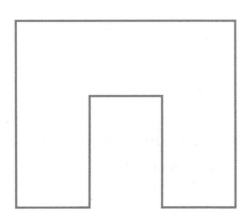

3 Try to find the perimeter of the complicated irregular shapes.

a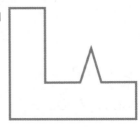

b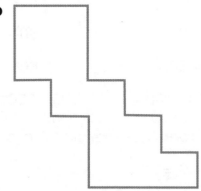

4 Draw shapes that have these perimeters on dot paper.

a 8 cm   **b** 12 cm   **c** 9 cm

56   Unit **14**   **Perimeter and Area** (TRB pp. 72–75)
Using units of measurement Compare objects using familiar metric units of area and volume
(ACMMG290) **AC**

Island Areas

You will need: BLM 17 '1 cm Grid Paper'

Captain Calculus has found some new islands on his travels. He needs to find their areas to see which is the largest for the new colony. Can you help him?

1 Find the area of each island.

a

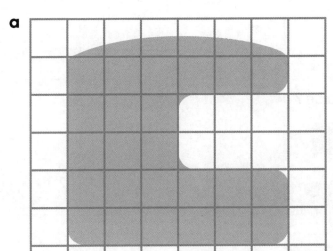

b

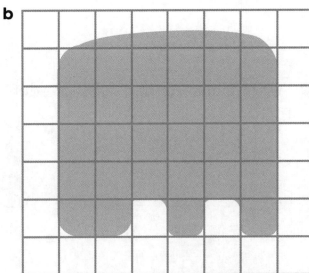

c

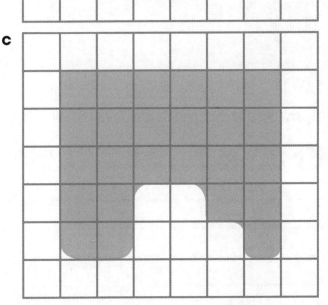

d

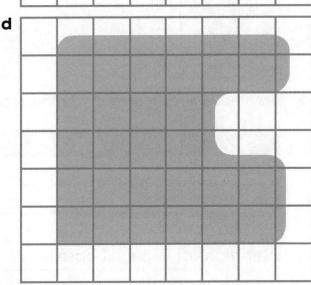

2 Rank the islands from **smallest** (1) to **largest** (4).

_____

3 Which is the best island for the new colony?

_____

Extension: Design your own island with an area of 50 cm² on grid paper.

Unit 14 **Perimeter and Area** (TRB pp. 72–75)
Using units of measurement Compare objects using familiar metric units of area and volume
(ACMMG290) **AC**

57

Finding Area

You will need: a ruler, BLM 17 '1 cm Grid Paper'

1 Find the area of each shape.

a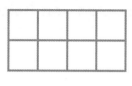

b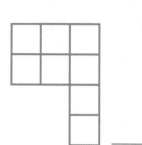

c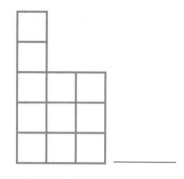

2 What is the shaded area of each shape?

a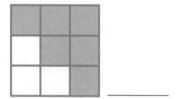

b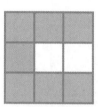

c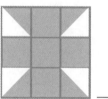

d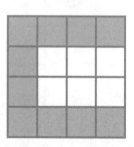

e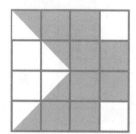

3 Tick the box to indicate whether each item would be measured in m² or cm².

Object	cm²	m²
a car park		
a slice of bread		
a playground		
the front of a pencil case		
a playing card		

4 Circle the **smaller** area in each pair.

a 27 cm² or 27 m²   **b** 41 cm² or 41 m²   **c** 3 m² or 3 cm²

5 Draw shapes that have these areas on grid paper.

a 20 cm²   **b** 32 cm²

58   Unit 14   **Perimeter and Area** (TRB pp. 72–75)
Using units of measurement Compare objects using familiar metric units of area and volume
(ACMMG290) AC

STUDENT ASSESSMENT

You will need: a ruler, BLM 17 '1 cm Grid Paper'

1 For each shape:

a find the **perimeter**.

b divide the shape into common shapes (e.g. squares and rectangles).

c find the total **area**.

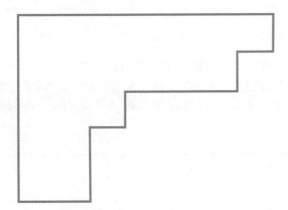

Perimeter:

Area:

Perimeter:

Area:

2 Circle the **larger** area in each pair.

a 41 cm² or 41 m²

b 50 cm² or 50 m²

c 65 cm² or 66 cm²

d 56 m² or 58 cm²

3 List 3 objects that would have an area greater than 1 m².

_____

4 List 3 objects that would have an area less than 1 m².

_____

Extension Use your grid paper to:

a draw an irregular object that has a perimeter of 30 cm.

b draw a compound object that has an area of 30 cm².

Perimeter and area (TRB pp. 72–75)
Using units of measurement Compare objects using familiar metric units of area and volume
(ACMMG290) **AC**

59

Commutative Arrays

1 **a** Draw the array for 4 × 2.

Remember: 4 × 2 means 4 rows of 2.

b Now draw the array for 2 × 4.

Remember: 2 × 4 means 2 rows of 4.

c What do you notice? _____

2 Complete the table, filling in the facts and drawing related arrays.

	Known fact	Array	Other fact	Other array
a	3 × 5 = 15			
b	5 × 2 =			
c	2 × 3 =			
d	1 × 7 =			
e	4 × 5 =			

3 On the multiplication grid, shade in all the tables that have reversed and equal equations.

4 What does the shading in the multiplication grid tell you about tables facts?

_____

_____

_____

×	1	2	3	4	5	6	7	8	9	10
1										
2										
3										
4										
5										
6										
7										
8										
9										
10										

Multiplication and Division Strategies (TRB pp. 76–79)
Number and place value Develop efficient mental and written strategies and use appropriate digital
technologies for multiplication and for division where there is no remainder
(ACMNA076) **AC**

Doubling and Halving

DATE:

1 Use doubles to complete the equations.

a double 4 =   2 × 4 =

b double 10 =   2 × 10 =

c double 7 =   2 × 7 =

d double 9 =   2 × 9 =

2 Use double doubles to complete the equations. One has been done.

Double		Double		Multiplication
10	10 + 10 = 20	20	20 + 20 = 40	4 × 10 =
6	=	12	=	4 × 6 =
5	=	10	=	4 × 5 =
12	=	24	=	4 × 12 =

3 Use halving to complete the equations.

a half of 18 =   18 ÷ 2 =

b half of 12 =   12 ÷ 2 =

c half of 8 =   8 ÷ 2 =

d half of 2 =   2 ÷ 2 =

4 Use halving twice to find the answers. One has been done.

Halve		Halve		Division
20	20 – 10 = 10	10	10 – 5 = 5	20 ÷ 4 =
40	=	20	=	40 ÷ 4 =
32	=	16	=	32 ÷ 4 =
48	=	24	=	48 ÷ 4 =

Extension: Use doubles to complete the equations.

a 8 × 25 =   **b** 16 × 21 =

Explain how you used doubles to help find the answer.

Unit 15 **Multiplication and Division Strategies** (TRB pp. 76–79)
Number and place value Develop efficient mental and written strategies and use appropriate digital
technologies for multiplication and for division where there is no remainder
(ACMNA076) **AC**

61

Hidden Arrays

1 Uncover the arrays to complete the multiplication equations.

a

4 × _____

b

4 × _____

c

_____

d

_____

e

_____

f

_____

Multiplication and Division Strategies (TRB pp. 76–79)
Number and place value Develop efficient mental and written strategies and use appropriate digital technologies for multiplication and for division where there is no remainder
(ACMNA076) **AC**

STUDENT ASSESSMENT

1 Fill in the matching reversed equivalent multiplication equation.

a 3 × 2 = ☐6☐ = 2 × _____   **b** 7 × 5 = ☐35☐ = _____

c 10 × 6 = _____   **d** 9 × 4 = _____

e 8 × 3 = _____   **f** 4 × 8 = _____

2 Complete the doubling to find the answers.

a double 2 =   **c** double 10 =

double 4 =   double 20 =

4 × 2 =   4 × 10 =

b double 6 =   **d** double 7 =

double 12 =   double 14 =

4 × 6 =   4 × 7 =

3 Complete each equation by halving.

a 16 ÷ 2 =   **b** 20 ÷ 2 =

c 24 ÷ 4 =   **d** 48 ÷ 4 =

e 80 ÷ 4 =   **f** 100 ÷ 4 =

4 State the multiplication equations that represent the arrays.

a   **b**   **c**

_____   _____   _____

Extension: Show the equation on a number line: 8 × 22

Unit
15
Multiplication and Division Strategies (TRB pp. 76–79)
Number and place value Develop efficient mental and written strategies and use appropriate digital
technologies for multiplication and for division where there is no remainder
(ACMNA076) AC

63

Exploring Arrays

1 Find all the related multiplication and division equations for each number.

a 14 _____

_____

b 32 _____

_____

c 60 _____

_____

d 40 _____

_____

2 Using the information from Question 1, list all the factors of each number.

a 14 _____

_____

b 32 _____

_____

c 60 _____

_____

d 40 _____

_____

3 True or false (T or F)?

a 2 is a factor of 11 _____

b 4 is a factor of 36 _____

c 9 is a factor of 45 _____

d 3 is a factor of 10 _____

4 Write a pair of factors for each number.

a 25 _____

b 48 _____

c 30 _____

d 27 _____

Extension: On a sheet of paper, draw a factor tree for: 60, 32, 14 and 40.

Unit **16** **More Multiplication and Division Strategies** (TRB pp. 80–83)
Number and place value Develop efficient mental and written strategies and use appropriate digital technologies for multiplication and for division where there is no remainder
(ACMNA076) **AC**

Multiplying by 10s and 100s

1 Fill in the values.

a 6 tens =   **b** 3 tens =

c 7 tens =   **d** 14 tens =

e 22 tens =   **f** 51 tens =

g 3 hundreds =   **h** 8 hundreds =

i 9 hundreds =   **j** 12 hundreds =

k 18 hundreds =   **l** 27 hundreds =

2 Complete the equations.

a $3 \times 10 =$ **b** $9 \times 10 =$

c $11 \times 10 =$ **d** $36 \times 10 =$

e $7 \times 100 =$ **f** $2 \times 100 =$

g $17 \times 100 =$ **h** $33 \times 100 =$

3 Complete the patterns.

a $2 \times 6 =$ **b** $3 \times 7 =$

$20 \times 6 =$ $30 \times 7 =$

$200 \times 6 =$ $300 \times 7 =$

c $5 \times 3 =$ **d** $7 \times 4 =$

$50 \times 3 =$ $70 \times 4 =$

$500 \times 3 =$ $700 \times 4 =$

4 Complete:

a 8×20 **b** 7×60

$= 8 \times 2 \times 10$ $= 7 \times 6 \times 10$

$=$ $=$

c 4×40 **d** 2×70

$= 4 \times 4 \times 10$ $= 2 \times 7 \times 10$

$=$ $=$

Extension: Complete the equations.

a $30 \times 12 =$ **b** $50 \times 11 =$ **c** $70 \times 10 =$

Unit 16 **More Multiplication and Division Strategies** (TRB pp. 80–83)
Number and place value Develop efficient mental and written strategies and use appropriate digital
technologies for multiplication and for division where there is no remainder
(ACMNA076) **AC**

65

Division Linked with Multiplication

1 Complete the division equations.

a $60 \div 10 =$ 　　　　　　　　**b** $4 \div 2 =$

c $25 \div 5 =$ 　　　　　　　　**d** $49 \div 7 =$

e $100 \div 10 =$ 　　　　　　　**f** $56 \div 7 =$

g $21 \div 3 =$ 　　　　　　　　**h** $36 \div 4 =$

i $12 \div 3 =$ 　　　　　　　　**j** $12 \div 6 =$

2 Complete the multiplication and division equations.

a $5 \times 7 =$ 　　　　　**b** $12 \times 4 =$ 　　　　　**c** $3 \times 8 =$

$7 \overline{)\ 35}$ 　　　　　　　$12 \overline{)\ 48}$ 　　　　　　　$3 \overline{)\ 24}$

$5 \overline{)\ 35}$ 　　　　　　　$4 \overline{)\ 48}$ 　　　　　　　$8 \overline{)\ 24}$

3 Fill in the missing numbers.

a _____ $\times 2 = 16$ 　　　**b** _____ $\times 4 = 36$ 　　　**c** $8 \times$ _____ $= 8$

d _____ $\times 10 = 80$ 　　**e** _____ $\times 6 = 36$ 　　**f** $3 \times$ _____ $= 15$

4 Solve the division problems.

a Divide 8 into 32. _____ 　　　**b** How many groups of 5 are in 45? _____

c How many are in each share if 28 are shared equally among 7? _____

5 **Riddle:** What is in seasons, seconds, centuries and minutes, but not in decades, years or days?

Look at all your answers to Questions 1–4. Cross out each number below that is the same as an answer above. Find the only number not crossed out; this corresponds to the letter in the alphabet that is the answer to the riddle.

7　　　5　　　9　　　2　　　4　　　14　　　1　　　3　　　6　　　8　　　10

Unit 16 **More Multiplication and Division Strategies** (TRB pp. 80–83)
Number and place value Develop efficient mental and written strategies and use appropriate digital technologies for multiplication and for division where there is no remainder
(ACMNA076) **AC**

DATE:

STUDENT ASSESSMENT

1 Look at the number 20.

a List all of the multiplication and division equations related to the answer 20.

_____

b Now list all of the factors of 20.

_____

2 Complete the multiplication equations.

a 3 × 10 =   **b** 6 × 10 =

c 13 × 10 =   **d** 34 × 10 =

Explain how you found the answer to **d**.

_____

3 Complete the patterns.

a 3 × 6 =   **b** 4 × 5 =

   30 × 6 =   40 × 5 =

   300 × 6 =   400 × 5 =

4 Complete the division equations.

a 60 ÷ 10 =   **b** 4 ÷ 2 =

c 25 ÷ 5 =   **d** 49 ÷ 7 =

e 9$\overline{)72}$ **f** 10$\overline{)50}$

g 3$\overline{)21}$ **h** 6$\overline{)36}$

Extension: Write one multiplication and one division fact for each triangle.

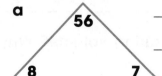

_____

_____

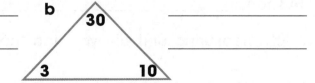

_____

_____

Unit
16

More Multiplication and Division Strategies (TRB pp. 80–83)
Number and place value Develop efficient mental and written strategies and use appropriate digital
technologies for multiplication and for division where there is no remainder
(ACMNA076) **AC**

67

Measuring Volume

You will need: 3 different-sized boxes, Centicubes, marbles, Unifix blocks, large counters, teddy counters, table tennis balls, MAB tens

1 Select a box and one of the measuring units above.

2 Record the measuring unit and how many units fit in the box in the table.

3 Repeat this 4 times, selecting a different measuring unit each time.

4 Move on to the second, and then third box, repeating steps 2–4 each time. Note that you can select different materials each time.

Measuring unit	Box 1	Box 2	Box 3
Centicubes			
marbles			
Unifix			
large counters			
teddy counters			
table tennis balls			
MAB tens			

5 List the advantages or disadvantages of each unit as a measure of volume.

Centicubes: _____

Marbles: _____

Unifix: _____

Large counters: _____

Teddy counters: _____

Table tennis balls: _____

MAB tens: _____

6 Which measuring unit did you find the best for measuring volume? Why?

_____

Cooper's Buildings

You will need: 12 Centicubes

Cooper used 12 blocks to build a building. What might Cooper's building look like?

Record your ideas on the page.

Hint: there is more than one answer!

Unit 17
Volume (TRB pp. 84–87)
Using units of measurement Compare objects using familiar metric units of area and volume
(ACMMG290) AC

69

Make the Volumes

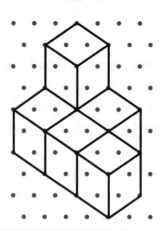

You will need: Centicubes

1 Make models using the number of Centicubes in the table.

2 Draw two different views of each model (from the front and from the side).

6 Centicubes	8 Centicubes
10 Centicubes	**9 Centicubes**
12 Centicubes	**This one is your choice! Volume:**

Unit **17** **Volume** (TRB pp. 84–87)
Using units of measurement Compare objects using familiar metric units of area and volume
(ACMMG290) **AC**

Unit 17

STUDENT ASSESSMENT

You will need: Centicubes

1 Imagine you want to measure the volume of a box.

a What would be the best measuring unit: tennis balls or blocks?

_____

b Explain your answer.

_____

2 Look at the diagram below.

a Make the model with Centicubes.

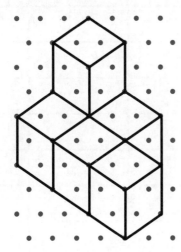

b What is the volume of this model? _____

c Remove 3 Centicubes from your model and draw two views of it above.

3 Compare the two models you created in Questions 2a and 2c.

a Which model has the greater volume? _____

b Explain your answer. _____

Unit 17 **Volume** (TRB pp. 84–87)
Using units of measurement Compare objects using familiar metric units of area and volume
(ACMMG290) **AC**

71

Place-Value Chart

1 Expand each decimal number by using the correct columns in the place-value chart.

a 42.39   **b** 21.78

c 130.45   **d** 5.54

e 17.09   **f** 12.6

g 710.56   **h** 11.08

i 403.6   **j** 0.56

Extension:

k five and six tenths

l seventeen and forty-two hundredths

m ninety and nine tenths

n twenty-seven hundredths

o nine hundred and twenty-five and three tenths

	hundreds	tens	ones	.	tenths	hundredths
a						
b						
c						
d						
e						
f						
g						
h						
i						
j						
k						
l						
m						
n						
o						

Unit 18 **Decimals to Two Decimal Places** (TRB pp. 88–91)
Fractions and decimals Recognise that the place-value system can be extended to tenths and hundredths. Make connections between fractions and decimal notation
(ACMNA079) **AC**

Matching Decimal Numbers and Words

1 Match the words and the decimal numbers. Write the letter that corresponds to the matching number in words on the line above each decimal number to complete the puzzle.

A	ninety-three and five hundredths
S	three and seven tenths
D	one hundred and six and thirty-nine hundredths
P	eighteen and twenty-five hundredths
O	eighty-one and seventy-six hundredths
E	eight and fifty-five hundredths
W	ninety-nine and ninety-nine hundredths
I	two hundred and sixteen and ninety-three hundredths
N	twenty-one and five hundredths
C	eighteen and five hundredths
T	three hundred and forty-two and thirty-nine hundredths
I	sixteen and thirty-nine hundredths
X	two hundred and fifteen
L	eight hundred and twenty-seven and twenty-five hundredths
M	four hundred and twenty-one and five hundredths
F	fifty-one and seven tenths
A	seventy-six and thirty-five hundredths

What is small and round?

A						
93.05						

106.39	8.55	18.05	216.93	421.05	76.35	827.25

18.25	81.76	16.39	21.05	342.39

Unit 18 **Decimals to Two Decimal Places** (TRB pp. 88–91)
Fractions and decimals Recognise that the place-value system can be extended to tenths and hundredths. Make connections between fractions and decimal notation
(ACMNA079) **AC**

73

Ordering Decimal Numbers

1 Order each set of decimal numbers from **smallest** to **largest**.

a   42.39   12.85   15.72   29.46   30.08

_____

b   20.36   25.49   21.39   19.86   23.76

_____

c   8.46   8.59   8.62   8.73   8.21

_____

d   150.26   150.73   150.07   150.54   150.37

_____

2 Create a number line for each set of decimals.

a   16.9   16.5   16.3   17.5   17.3

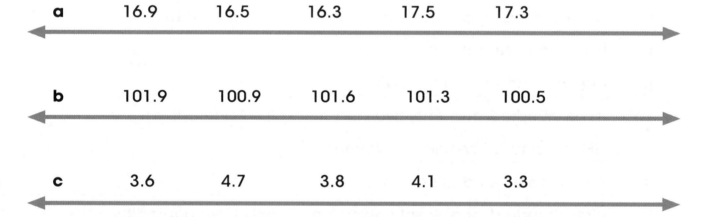

b   101.9   100.9   101.6   101.3   100.5

c   3.6   4.7   3.8   4.1   3.3

3 These times in minutes are from a car race. Place them in order from **shortest** to **longest** to show 1st, 2nd and 3rd.

9.45   9.00   8.45   9.30   9.15   9.27   9.06   8.54

_____

4 These are the weights in kilograms of 5 elephants. Place them in order from **largest** to **smallest** to find the heaviest and the lightest elephant.

3 200.49   5 505.69   4 370.56   5 505.05   4 465.98

_____

Unit 18 **Decimals to Two Decimal places** (TRB pp. 88–91)
Fractions and decimals Recognise that the place-value system can be extended to tenths and
hundredths. Make connections between fractions and decimal notation
(ACMNA079) **AC**

Unit **18**

STUDENT ASSESSMENT

Look at these numbers:

| 2.45 | 3.07 | 4.6 | 3.90 | 3 | 2.42 | 4.05 |

1 Order the numbers from **smallest** to **largest**.

_____

2 Select one of the numbers that has 2 decimal places.

Write it here: _____ This is your special number.

3 Write your special number in words.

_____

4 Include all of the numbers in the box on the number line below.

⟵——————————————————————————————⟶

5 Explain where you placed your special number on the number line and why. _____

6 Is the number 5.01 **smaller** or **larger** than your special number?

_____

7 How do you know? _____

_____

8 Draw a picture of your special number.

Unit **18** **Decimals to Two Decimal Places** (TRB pp. 88–91)
Fractions and decimals Recognise that the place-value system can be extended to tenths and hundredths. Make connections between fractions and decimal notation
(ACMNA079) **AC**

75

Chance in Everyday Life

1 Choose the word that **best** describes each statement.

> least likely   unlikely   likely   equal chance   most likely

a It will be raining tomorrow. _____

b If I drop a glass it will break. _____

c If a coin is tossed it will be a tail. _____

d The sun will set tonight. _____

e I will go to school tomorrow. _____

2 Choose the word that **best** describes the chance of landing on the shaded area of each spinner.

> least likely   unlikely   likely   equal chance   most likely

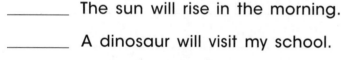

a _____   **b** _____   **c** _____

3 Order the events from **least likely** to **most likely** to occur.
1 is the **least likely** and 6 is the **most likely**.

_____ The sun will rise in the morning.

_____ A dinosaur will visit my school.

_____ A new baby will be a boy.

_____ I will have dinner at 6 pm today.

_____ I will see a bird on the way home from school.

_____ I will roll a 4 on a dice.

Extension: Write 3 examples of events that are **most likely** to happen.

_____

_____

What's the Chance?

You will need: a coin

Try the activities.

1 A coin is tossed randomly onto a set of parallel lines.
What is the chance the coin touches a line?

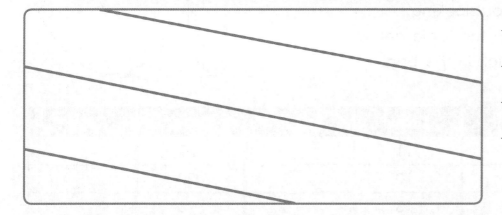

_____

_____

_____

_____

2 A coin is tossed randomly onto a set of wider parallel lines.
What is the chance the coin touches a line?

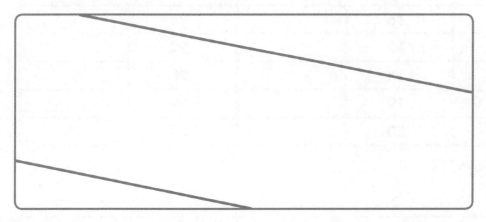

_____

_____

_____

_____

Make sure you can
explain your answer!

3 The coin is now tossed onto a circle with circular lines.
What is the chance the coin touches a line?

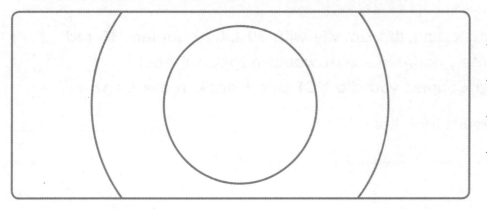

_____

_____

_____

_____

Unit **19** **Chance** (TRB pp. 92–95)
Chance Identify everyday events where one cannot happen if the other happens
(ACMSP093) AC

77

Five Counters in a Bag

You will need: 3 blue counters, 1 red counter, 1 green counter, a cloth bag

1 Put the counters in the bag.
Draw one out without looking.
Record the result in the table below.
Put the counter back in the bag.
Repeat 30 times.

Draw	Result	Draw	Result	Draw	Result
1		11		21	
2		12		22	
3		13		23	
4		14		24	
5		15		25	
6		16		26	
7		17		27	
8		18		28	
9		19		29	
10		20		30	

2 What have you found?

_____

3 Was draw number 8 affected by the previous draws? Explain your answer.

_____

Extension: Think about starting this activity with 30 blue counters, 10 red
counters and 10 green counters. What would happen if each
time you took out a counter, you did NOT put it back in the bag?

What might your results look like? _____

_____

Try this activity.

Chance (TRB pp. 92–95)
Chance Identify events where the chance of one will not be affected by the occurrence of the other
(ACMSP094) **AC**

STUDENT ASSESSMENT

1 Choose the word that **best** describes each statement.

> least likely   unlikely   likely   equal chance   most likely

a I will go to school tomorrow. _____

b If a dice is rolled, it will be a 4. _____

c If a coin is tossed, it will be a head. _____

d The sun will rise tomorrow. _____

e I will see a caterpillar at school this week. _____

2 Choose the word that **best** describes the chance of landing on the shaded area of each spinner.

> least likely   unlikely   likely   equal chance   most likely

a

b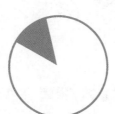

c

_____ _____ _____

3 Give an example of two events where one cannot happen if the other happens. _____

_____

4 **a** What is the chance of a baby bird being male or female, if the last bird hatched was male? _____

b Explain your answer.

_____

_____

Unit 19 **Chance** (TRB pp. 92–95)
Chance Describe possible everyday events and order their chances of occurring
(ACMSP092) **AC**

Identify everyday events where one cannot happen if the other happens
(ACMSP093) **AC**

Identify events where the chance of one will not be affected by the occurrence of the other
(ACMSP094) **AC**

79

Daisy's Data

Poor Daisy dropped all of the data that she collected. Can you help her organise it into the two different surveys she was conducting, using the table below? One was on animals and the other was on sport. You may organise the data any way you like. Hint: the numbers show how many of each item.

1   Daisy's data:

What is your favourite animal?	What is your least favourite sport?

2   What were two interesting things about each of Daisy's surveys?

_____

Unit **20**   **Collecting Data** (TRB pp. 96–99)
Data representation and interpretation Select and trial methods for data collection, including survey questions and recording sheets
(ACMSP095) **AC**

Venn Diagrams and 2-Way Tables

1   Place the data into the 2-way table.

- Toby has a bike and a scooter.
- Eliza doesn't have a scooter but has a bike.
- Connor has a bike but not a scooter.
- Grace doesn't have a bike or a scooter.
- Alex has a bike and a scooter.
- Gemma has a scooter but not a bike.

- Drew has a bike but not a scooter.
- Arthur has both.
- Lily has neither a bike nor a scooter.
- Justin has a bike and scooter.
- Dolores hasn't got a bike or a scooter.
- Damon has a bike and a scooter.

	Has a bike	Doesn't have a bike
Has a scooter		
Doesn't have a scooter		

2   Transfer this information into the Venn diagram.

3   Write an interesting fact from the above data.

_____

_____

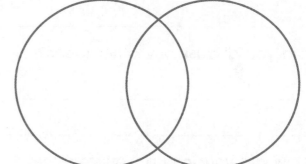

4   Here is a different Venn diagram.

   a How many players are shown? _____

   b How many players prefer serving? _____

   c How many players like serving and smashing? _____

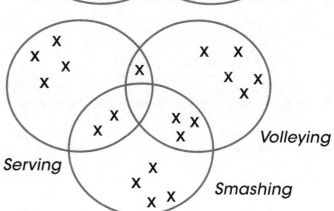

Serving

Volleying

Smashing

Unit 20   **Collecting Data** (TRB pp. 96–99)
Data representation and interpretation Select and trial methods for data collection, including survey questions and recording sheets
(ACMSP095) **AC**

81

Our Question: My Report

1 Record your question here.

_____

_____

2 What have you learned about the question so far?

_____

_____

_____

3 How did your group decide to present the information?

_____

_____

_____

4 How was that decision made?

_____

_____

5 Did you discuss any other ideas?

_____

_____

6 How did your group work together?

_____

_____

7 What might you do differently, if you had to complete the same task again?

_____

_____

_____

1 Look at the Venn diagram. Write 3 statements that can be observed from the information.

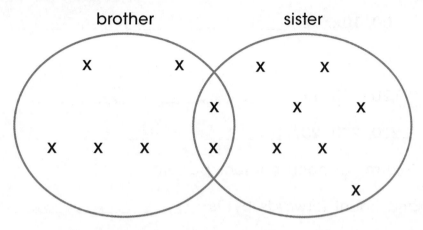

_____

_____

2 Look at the 2-way table. Write 3 statements that can be observed from the information.

	Has a dog	Doesn't have a dog
Has a cat	x x x x x	x x x
Doesn't have a cat	x x	x x x x x x x

_____

_____

_____

_____

3 a Write a survey question that could be used to ask people about their favourite eating location.

_____

b Collect the data by asking 5 different people your question.

Extension: On a sheet of paper, create a table, a Venn diagram or a 2-way table of your data.

Unit
20 **Collecting Data** (TRB pp. 96–99)
Data representation and interpretation Select and trial methods for data
collection, including survey questions and recording sheets
(ACMSP095) **AC**

83

Number Patterns

1 Continue the number patterns.

a + 4, – 2   1, 5, 3, ____, ____, ____, ____, ____, ____, ____, ____, ____

b + 2, – 10   100, 102, 92, ____, ____, ____, ____, ____, ____, ____, ____, ____

c + 5, – 8   70, 75, 67, ____, ____, ____, ____, ____, ____, ____, ____, ____

d + 10, – 6   120, 130, 124, ____, ____, ____, ____, ____, ____, ____, ____, ____

e + 20, – 13   200, 220, 207, ____, ____, ____, ____, ____, ____, ____, ____, ____

2 Write the first 5 terms of each number pattern.

a Start at 90 and count forwards by 9s. ____, ____, ____, ____, ____

b Start at 60 and count backwards by 3s. ____, ____, ____, ____, ____

c Start at 200 and count backwards by 20s. ____, ____, ____, ____, ____

d Start at 40 and count forwards by 4s. ____, ____, ____, ____, ____

3 Describe each number pattern.

a 12, 20, 28, 32, 40, 48 _____

b 106, 112, 118, 124, 130 _____

c 1 000, 990, 980, 970, 960 _____

4 Complete the number patterns on the ladders.

a

| 110 |
| 65 |
| 50 |

b

| 60 |
| 24 |

c

| 105 |
| 85 |
| 65 |

Extension: Start at 200 and count backwards by 9s to find the 10th term.

_____

Unit 21 **Number Patterns** (TRB pp. 100–103)
Patterns and algebra Explore and describe number patterns resulting from performing multiplication
(ACMNA081) **AC**

Exploring Tables Patterns

You will need: BLM 17 '1 cm Grid Paper'

1 Look at the patterns below, made on a 100 chart.

a 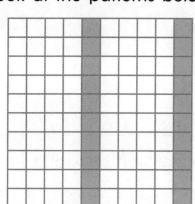 **b** 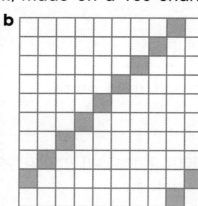 **c**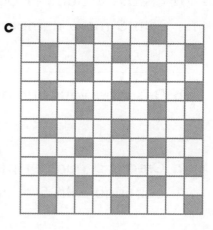

Can you identify the tables that made these patterns?

a _____   b _____   c _____

2 Look at the grids and patterns below. The charts are different sizes (not 10 × 10).

a 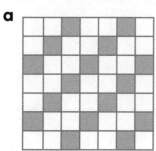 **b** 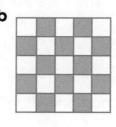 **c** 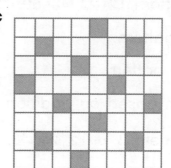 **d**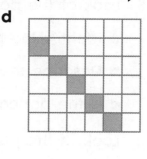

What tables could have made these patterns? Can you think why they
made them like that? Use grid paper to help you investigate.

a _____   b _____

c _____   d _____

Extension: Look at the centres of some grid patterns. On a sheet of paper,
identify the tables that created these patterns. (Hint: there might be
more than one answer!)

a 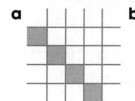 **b** 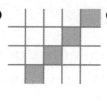 **c** 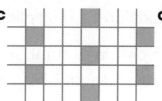 **d** 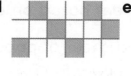 **e**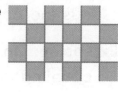

Unit 21 **Number Patterns** (TRB pp. 100–103)
Patterns and algebra Explore and describe number patterns resulting from performing multiplication
(ACMNA081) **AC**

85

More Number Patterns

You will need: a calculator

1 a Create a number pattern based on addition. Start at 4 and continue with 5 more numbers.

_____

b Describe your number pattern in words.

_____

2 a Create a number pattern based on multiplication. Start at 6, and continue with 5 more numbers.

_____

b Describe your number pattern in words.

_____

3 Look at the pattern 12, 24, 48, 60, 72.

a Identify the pattern.

b Describe the pattern. _____

c Is the pattern based on addition or multiplication? _____

4 Look at the pattern 3, 6, 12, 24, 48, 96.

a Identify the pattern.

b Describe the pattern. _____

c Is the pattern based on addition or multiplication? _____

5 Look at the pattern 5, 15, 45, 135, 405.

a Identify the pattern.

b Describe the pattern. _____

c Is the pattern based on addition or multiplication? _____

Extension: On a sheet of paper, create 2 number patterns starting at 100 and counting backwards. One pattern should be based on subtraction, and one on division.

DATE:

STUDENT ASSESSMENT

1 Continue the number patterns.

a + 5, – 1   1, 6, 5, ___, ___, ___, ___, ___, ___, ___, ___, ___

b + 20, – 5   100, 120, 115, ___, ___, ___, ___, ___, ___, ___, ___

c – 5, + 6   70, 65, 71, ___, ___, ___, ___, ___, ___, ___, ___, ___

2 Write the first 5 terms of each number pattern.

a Start at 50 and count forwards by 3s. ___, ___, ___, ___, ___

b Start at 100 and count backwards by 4s. ___, ___, ___, ___, ___

3 Describe each number pattern.

a 13, 21, 29, 37 _____

b 501, 491, 481, 471 _____

4 **a** Look at the 100 chart below (left) and describe the number pattern represented on it. _____

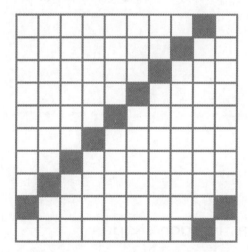 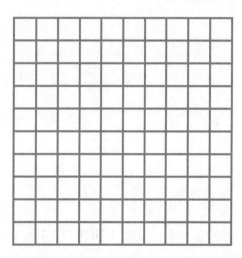

b Create your own pattern on the 100 chart above (right) and describe it.

_____

5 List 3 examples of where you can find number patterns in real life.

_____

_____

_____

Unit **21**   **Number Patterns** (TRB pp. 100–103)
Patterns and algebra Explore and describe number patterns resulting from performing multiplication
(ACMNA081) **AC**

87

Right Angles

1 Circle the right angles.

a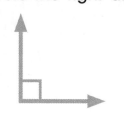
b
c

d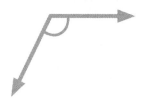
e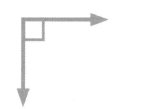
f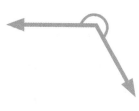

g
h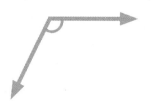
i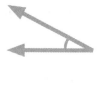

2 Using the starting lines, complete the right angles.

a
b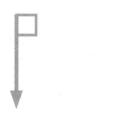
c

d
e
f

3 Show using a ▢ the right angle or angles in each shape.

a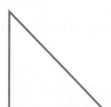
b
c
d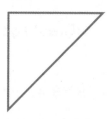

Extension: On a sheet of paper, write the letters of the alphabet in capitals. How many right angles did you find?

Angles (TRB pp. 104–107)
Geometric reasoning Compare angles and classify them as equal to, greater than or less than a right angle
(ACMMG089) **AC**

Acute Angles

1 Circle the acute angles.

a 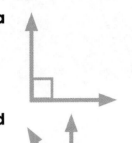 b  c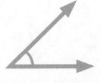

d  e  f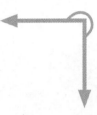

g  h   i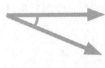

2 Using the starting lines, create acute angles.

a  b  c

d   e  f

3 What is an acute angle? _____

4 Order the angles from **smallest** (1) to **largest** (6).

a  b 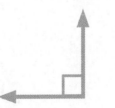 c

d   e  f

Unit **22** **Angles** (TRB pp. 104–107)
Geometric reasoning Compare angles and classify them as equal to, greater than or less than a right angle
(ACMMG089) **AC**

89

Comparing Angles

1 Circle the obtuse angles.

a 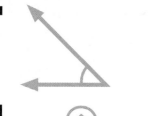   b 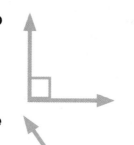   c

d    e    f

2 Order each set of angles from **smallest** (1) to **largest** (4).

Set 1 a   b    c 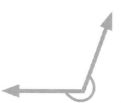   d

Set 2 a    b   c    d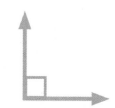

3 Circle the **larger** angle in each pair.

a   b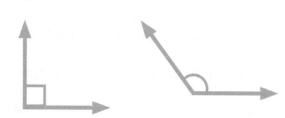

4 Draw an angle

a
smaller than:

b
larger than:

Extension: On a sheet of paper, draw a picture of a rocket. Label the acute angles (a), obtuse angles (o) and right angles (☐).

Unit **22** **Angles** (TRB pp. 104–107)
Geometric reasoning Compare angles and classify them as equal to, greater than or less than a right angle
(ACMMG089) **AC**

Unit
22

DATE:

STUDENT ASSESSMENT

You will need: a protractor and a ruler

1 Complete the table.

		Diagram of angle	
a	right angle		
b	angle less than a right angle		This angle is called:
c	angle greater than a right angle		This angle is called:

2 Circle the **larger** angle in each pair.

a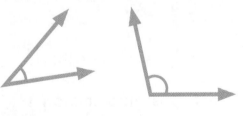

b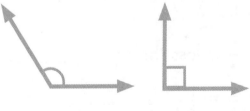

c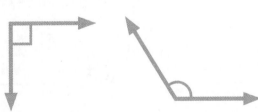

d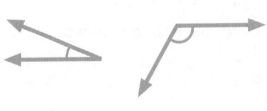

Extension: Draw the angles, using a protractor and a ruler.

a 60°   **b** 40°   **c** 145°

Angles (TRB pp. 104–107)
Geometric reasoning Compare angles and classify them as equal to, greater than or less than a right angle
(ACMMG089) **AC**

Halves, Quarters and Eighths

1 Shade part of each shape to match the given fraction.

a $\frac{1}{2}$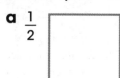

b $\frac{1}{4}$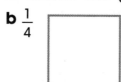

c $\frac{3}{4}$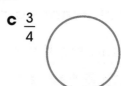

d $\frac{3}{8}$

e $\frac{5}{8}$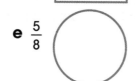

f $\frac{1}{8}$

2 What fraction of each shape has been shaded?

a 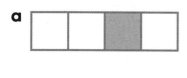 _____

b  _____

c 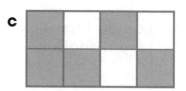 _____

d 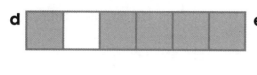 _____

e _____

f 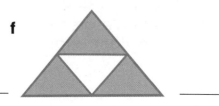 _____

3 Write true or false (T or F) for each statement.

a $1 = \frac{4}{4}$ _____

b $\frac{5}{8} = 1$ _____

c $\frac{1}{4}$ is less than $\frac{2}{8}$ _____

d $\frac{5}{8}$ is greater than $\frac{1}{2}$ _____

e $\frac{2}{4} = \frac{1}{2}$ _____

f $\frac{3}{4}$ is greater than 1 _____

4 Write the fractions in words.

a $\frac{3}{4}$ _____

b $\frac{7}{8}$ _____

c $\frac{1}{2}$ _____

5 Write the correct number in each box to find the equivalent fractions.

a $\frac{4}{8} = \frac{\boxed{}}{2}$

b $\frac{6}{8} = \frac{\boxed{}}{4}$

c $\frac{2}{8} = \frac{\boxed{}}{4}$

d $\frac{2}{4} = \frac{\boxed{}}{2}$

Extension: Which is larger: three quarters or seven eighths? On a sheet of paper, draw a diagram to support your answer.

Fraction Strips

1 Shade part of each fraction strip to match the given fraction.

a $\frac{1}{2}$

b $\frac{2}{3}$

c $\frac{5}{8}$

d $\frac{7}{10}$

2 What fraction of each fraction strip has been shaded?

a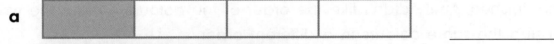

b

c

d

3 Shade each fraction strip to make the fractions equivalent.

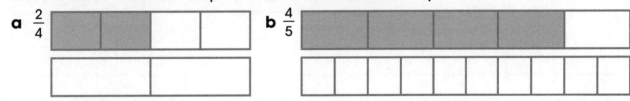

a $\frac{2}{4}$ **b** $\frac{4}{5}$

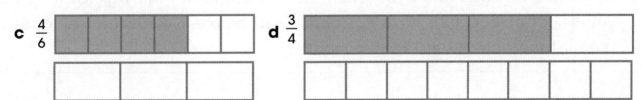

c $\frac{4}{6}$ **d** $\frac{3}{4}$

4 Write the fractions in words.

a $\frac{1}{6}$ _____   **b** $\frac{2}{3}$ _____

5 Write the correct number in each box to find the equivalent fractions.

a $\frac{4}{6} = \frac{\boxed{}}{3}$ **b** $\frac{2}{\boxed{}} = \frac{1}{3}$ **c** $\frac{1}{2} = \frac{\boxed{}}{6}$

Unit 23 **Equivalent Fractions** (TRB pp. 108–111)
Fractions and decimals Investigate equivalent fractions used in contexts
(ACMNA077) **AC**

93

Painting a Fraction Wall

You will need: BLM 40 'Fraction Wall', coloured pencils

Amity has been asked to paint the wall. She can only reach the first 3 rows from the ground.

1 Use your coloured pencils to colour Amity's wall.

$\frac{1}{10}$ red $\frac{4}{10}$ blue $\frac{3}{10}$ yellow $\frac{2}{10}$ green

2 After she finished, Amity didn't like the order of the colours. Colour the wall again, using the same colours in a different order.

3 Amity then painted the next section of the wall.

$\frac{2}{5}$ pink $\frac{1}{5}$ purple $\frac{2}{5}$ red

4 Amity has decided to let you paint the last section of the wall, as long as you select colours that have not been used before. Colour the wall and then list the colours and the fractions.

Extension: Can you work out what fraction of the sections of the walls in Questions 2 and 3 was painted red?

_____

_____

  Equivalent Fractions (TRB pp. 108–111)
Fractions and decimals Investigate equivalent fractions used in contexts
(ACMNA077) AC

Unit 23 STUDENT ASSESSMENT

1 Shade each of the shapes so that they represent $\frac{1}{2}$.

a 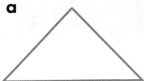 **b** **c** 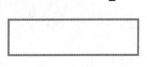 **d**

2 Explain what an equivalent fraction is.

_____

_____

3 List 3 equivalent fractions.

_____

4 Draw a diagram of each fraction from Question 3.

5 Work with the fraction strips below.

a Divide one into tenths and one into fifths.

b Colour $\frac{4}{10}$ red and $\frac{2}{5}$ blue.

c What do you notice? _____

_____

Unit 23 **Equivalent Fractions** (TRB pp. 108–111)
Fractions and decimals Investigate equivalent fractions used in contexts
(ACMNA077) **AC**

95

Improper Fractions to Mixed Numbers

1 Circle the improper fractions.

a $\frac{3}{4}$ **b** $1\frac{1}{2}$ **c** $\frac{5}{3}$ **d** $2\frac{1}{4}$ **e** $\frac{8}{5}$ **f** $\frac{1}{5}$

2 How did you identify the improper fractions?

_____

_____

3 Change the improper fractions to mixed numbers.

a $\frac{8}{5}$ =   **b** $\frac{3}{2}$ =   **c** $\frac{10}{3}$ =

d $\frac{5}{4}$ =   **e** $\frac{10}{8}$ =   **f** $\frac{11}{6}$ =

4 Order each fraction set from **smallest** to **largest**.

a $\frac{1}{2}$ $\frac{2}{3}$ $\frac{1}{3}$ 1   2 $\frac{4}{3}$ _____

b $\frac{3}{5}$ $\frac{2}{10}$ 1 $\frac{11}{10}$ 2 $1\frac{1}{2}$ _____

5 Draw a number line to show a number sequence starting at $1\frac{1}{2}$ and counting by $\frac{1}{2}$ to 5.

Extension: Order the fractions from **smallest** to **largest**.

$\frac{2}{4}$ $\frac{1}{8}$ $\frac{1}{2}$ 2 $\frac{5}{8}$ $\frac{5}{4}$ $\frac{8}{8}$

96   **Unit 24**   **Counting with Fractions** (TRB pp. 112–115)
Fractions and decimals Count by quarters, halves and thirds, including with mixed numerals.
Locate and represent these fractions on a number line
(ACMNA078) **AC**

Mixed Numbers to Improper Fractions

1 Colour the fractions in the grid that are mixed numbers.

$1\frac{1}{2}$	$\frac{4}{5}$	$\frac{8}{5}$	$\frac{4}{5}$
$\frac{7}{8}$	$2\frac{1}{8}$	$\frac{1}{6}$	$\frac{9}{4}$
$\frac{10}{3}$	$3\frac{1}{5}$	$2\frac{1}{4}$	$\frac{11}{10}$

2 Change the mixed numbers to improper fractions.

a $1\frac{1}{4} =$ **b** $2\frac{1}{2} =$ **c** $3\frac{1}{3} =$

d $2\frac{5}{8} =$ **e** $3\frac{4}{5} =$ **f** $1\frac{7}{8} =$

3 Order each fraction set from **smallest** to **largest**.

a $3\frac{1}{4}$ $2\frac{1}{2}$ $1\frac{1}{4}$ 2 $2\frac{1}{5}$ _____

b $4\frac{1}{8}$ $4\frac{1}{2}$ 3 $2\frac{1}{4}$ $1\frac{1}{2}$ _____

4 Change the mixed numbers to improper fractions, and then order from **smallest** to **largest**.

a $3\frac{1}{2} =$ $\frac{4}{2} =$ $1\frac{1}{2} =$ $\frac{5}{2} =$ $\frac{9}{2} =$

_____

b $\frac{3}{8}$ $1\frac{1}{4} =$ $\frac{17}{8}$ 2 $1\frac{5}{8} =$

_____

Extension: Complete a number line for the fractions from Question 4.

Unit 24 **Counting with Fractions** (TRB pp. 112–115)
Fractions and decimals Count by quarters, halves and thirds, including with mixed numerals.
Locate and represent these fractions on a number line
(ACMNA078) **AC**

97

Ordering Fractions

1 Order each fraction set from **smallest** to **largest**.

a $\frac{1}{4}$ $\frac{1}{2}$ $\frac{1}{3}$ $\frac{2}{3}$ $\frac{3}{4}$ _____

b $\frac{1}{10}$ $\frac{4}{5}$ $\frac{3}{10}$ 1 $\frac{1}{2}$ _____

c $\frac{1}{8}$ $1\frac{1}{4}$ $\frac{3}{4}$ 1 $\frac{5}{8}$ _____

2 Complete each statement with < or > or =.

a $\frac{1}{2}$ $\frac{2}{3}$ b $\frac{3}{4}$ $\frac{4}{8}$ c $\frac{2}{3}$ $\frac{1}{2}$

d $\frac{4}{5}$ 1   e $\frac{5}{8}$ $\frac{3}{4}$ f $\frac{7}{10}$ $\frac{3}{5}$

3 Circle the **larger** fraction in each pair.

a $1\frac{1}{2}$ $\frac{5}{2}$ b $\frac{3}{4}$ $1\frac{1}{8}$ c $\frac{10}{3}$ 3

d $\frac{8}{5}$ $2\frac{1}{4}$ e $3\frac{1}{5}$ $\frac{10}{5}$ f $\frac{5}{2}$ $2\frac{1}{4}$

4 Complete a number line for each set of fractions.

a $\frac{3}{2}$ $\frac{1}{4}$ 1 $\frac{3}{4}$ $\frac{2}{4}$

b $\frac{5}{3}$ 1 $1\frac{1}{3}$ $\frac{2}{3}$ $\frac{10}{3}$

Extension: Which is larger: one quarter or one third? Draw a diagram to support your answer.

Counting with Fractions (TRB pp. 112–115)
Fractions and decimals Count by quarters, halves and thirds, including with mixed numerals.
Locate and represent these fractions on a number line
(ACMNA078) **AC**

DATE:

STUDENT ASSESSMENT

Unit 24

1 **a** Circle the improper fractions below.

b Put stars next to the mixed numbers below.

$1\frac{1}{4}$ $\frac{3}{2}$ $\frac{1}{4}$ $\frac{3}{5}$ $\frac{4}{10}$

$\frac{2}{5}$ $7\frac{1}{2}$ $\frac{8}{3}$ $2\frac{1}{4}$ $\frac{10}{3}$

$\frac{6}{5}$ $5\frac{1}{5}$ $\frac{1}{5}$ $\frac{8}{5}$ $3\frac{1}{6}$

2 Explain what a mixed number is.

_____

_____

3 Change the improper fractions to mixed numbers.

a $\frac{7}{3}$ =   **b** $\frac{10}{4}$ =   **c** $\frac{20}{6}$ =

4 Change the mixed numbers to improper fractions.

a $3\frac{1}{4}$ =   **b** $4\frac{1}{2}$ =   **c** $8\frac{3}{10}$ =

5 Order the fractions from **smallest** to **largest**.

$\frac{7}{2}$ 3 $2\frac{1}{2}$ $\frac{4}{2}$ $3\frac{1}{4}$

Extension: Draw a number line showing a number sequence starting at 0 and counting by quarters.

Unit 24
Counting with Fractions (TRB pp. 112–115)
Fractions and decimals Count by quarters, halves and thirds, including with mixed numerals.
Locate and represent these fractions on a number line
(ACMNA078) **AC**

99

Time

1 Complete the clocks to show the times.

a half past 3

b quarter to 9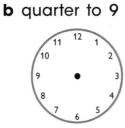

c quarter past 11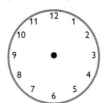

d half past 8

e quarter past 1

f quarter to 6

2 Write the time shown on each clock.

a

b

c

_____   _____   _____

3 Complete the labels for the digital times.

a

b

c

| : |

| : |

| : |

4 Draw a line matching the digital and analogue clocks.

a

b

c

| **4 : 45** |

| **3 : 15** |

| **1 : 00** |

Unit 25

Time (TRB pp. 116–119)
Using units of measurement Convert between units of time
(ACMMG085) **AC**

Converting Between Units of Time

You will need: a calculator

1 Complete the conversions.

   a 60 minutes = _____ hour   **b** 1 year = _____ months

   c _____ hours = 1 day   **d** _____ hours = 180 minutes

   e _____ seconds = 2 minutes   **f** _____ minutes = 600 seconds

2 Match each length of time with a word from the word list.

> seconds   minutes   hours   days   weeks   months   years

   a a football match _____   **b** school holidays _____

   c the age of a town _____   **d** the length of a song _____

3 Use the symbols < or > to make the statements true. Remember the largest part of the sign is "eating" the **greatest** value.

   a 70 minutes is _____ 1 hour   **b** 25 days is _____ 1 month

   c 65 seconds is _____ 1 minute   **d** 20 hours is _____ 1 day

   e 400 days is _____ 1 year   **f** 6 weeks is _____ 1 month

4 Name the month that comes before:

   a June _____   **b** October _____

   c December _____   **d** May _____

   e July _____   **f** February _____

Extension: Find your age in years, days, minutes and, if you are brave, seconds!

Unit **25**   **Time** (TRB pp. 116–119)
Using units of measurement Convert between units of time
(ACMMG085) **AC**

101

am and pm Time

Match the times written below and write down the corresponding letters.

O 4 am   **S** 12:55 am

A 10 am   **D** 3 am

T 8 am   **E** 9:25 pm

Y 9:30 am   **M** 1:45 pm

N 7:15 pm   **I** 1 am

What did the White Rabbit say?

   ten o'clock in the morning      **4 : 00**

_____ _____ _____ _____

         **09 : 25**

_____ _____ _____ _____

   **4 : 00**      ten o'clock in the morning

_____ _____ _____ _____

STUDENT ASSESSMENT

You will need: a calculator

1 Write the time shown on each clock.

a    **b**    **c**    **d**

_____   _____   _____   _____

2 Complete the clocks to show the times.

a quarter to 6   **b** quarter past 7   **c** half past 1   **d** 9 o'clock

   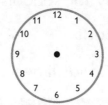   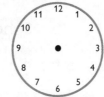   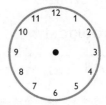

3 Complete the conversions.

a 60 minutes = _____ hour   **b** 24 hours = _____ day

c 1 year = _____ days   **d** _____ months = 1 year

4 Circle the **larger** amount of time in each pair.

a   90 minutes   1 hour

b   15 days   2 weeks

c   6 weeks   1 month

d   240 minutes   5 hours

5 Complete the clocks with the missing times.

a | **4 : 15** | **b** | **:** | **c** | **12 : 45** | **d** | **:** |

        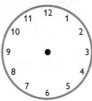   

Ashleigh's Day

1 Write the times in the correct order. Write each time on the digital clock.

nine o'clock in the morning	7:30 am	
ten-thirty in the morning	1:00 pm	11 in the morning
twenty-five past eight at night	6:30 pm	6:45 am
7:00 am	3:35pm	half-past three

2 Match the daily events to the times on the digital clocks.

Dinner with the family	School finishes	Wake up
Lunchtime	Get ready for school	After-school footy
Breakfast	School starts	Bedtime

3 Add two more activities for the additional times.

Event	Digital time
	:
	:
	:
	:
	:
	:
	:
	:
	:
	:
	:

Time Problems (TRB pp. 120–123)
Using units of measurement Use am and pm notation and solve simple time problems
(ACMMG086) **AC**

Adelaide to Darwin

You will need: a calculator

This travel line shows the distances between Adelaide and Darwin. (The line is not to scale.) Use the travel line to answer the questions. Show your working.

1   How long does the trip take in days, hours and minutes?

2   How far is it from Cooper Pedy to Alice Springs? How long does it take?

3   What is the distance between Tennant Creek and Katherine? How long does it take?

4   What is the shortest leg of the trip?

5   What is the longest length of time between two locations?

6   Which town would be about halfway? Explain how you worked out your answer.

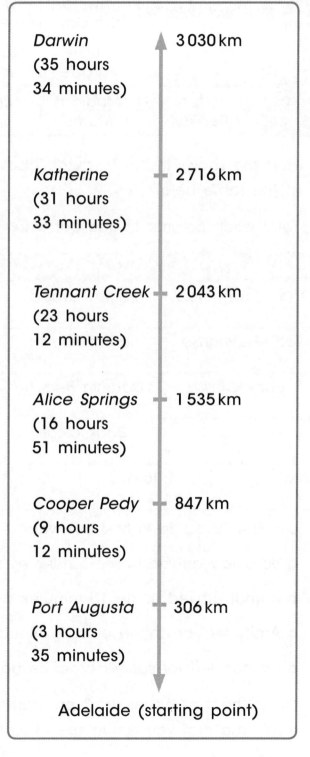

Darwin
(35 hours
34 minutes)   3030 km

Katherine
(31 hours
33 minutes)   2716 km

Tennant Creek   2043 km
(23 hours
12 minutes)

Alice Springs   1535 km
(16 hours
51 minutes)

Cooper Pedy   847 km
(9 hours
12 minutes)

Port Augusta   306 km
(3 hours
35 minutes)

Adelaide (starting point)

Unit 26 **Time Problems** (TRB pp. 120–123)
Using units of measurement Use am and pm notation and solve simple time problems
(ACMMG086) **AC**

105

Television Time

	3:00 pm	3:30 pm	4:00 pm	4:30 pm	5:00 pm	5:30 pm
TV 1	The Numbers Show	The Clown Club	Ted's Playhouse	Safari Games	Dolphin Beach	Book Chat
TV 2	Weird Wildlife	News of the Day	Travel Adventures	Football	The Cube	Library Detectives
TV 3		Puffin Parade	Laugh it Off	News of the Day		Quantum Quiz
TV 4	Virtual Reality!	Magical Maths	Bake Club	News of the Day		Front Seat

1 Find the shows in the TV guide above. Record how long each show ran for in the table below.

2 Total each column to find out who watched the most TV.

Veronica	Jamal	Amity	Damon
The Clown Club	Puffin Parade	Magical Maths	Weird Wildlife
Ted's Playhouse	Safari Games	Front Seat	Football
Book Chat	Dolphin Beach		
	Book Chat		
Total:	Total:	Total:	Total:

3 Use the TV guide to find out when each child got home or left.

a Veronica arrived home 5 minutes before *The Clown Club* started. _____

b Jamal arrived home 15 minutes after *Weird Wildlife* started. _____

c Amity left for dance class 5 minutes after the end of *Front Seat*. _____

d Damon left for soccer practice halfway through *Library Detectives*. _____

4 On a sheet of paper, list 3 programs that you would watch, and find the total time that you would spend watching TV.

Unit 26 STUDENT ASSESSMENT

1 Order the times from **earliest** (1) to **latest** (6).

7 am   03:00   7:30 in the evening   13:30   9 pm   11:00 in the morning

_____

2 Add times to complete the timetable. Make sure you write if they are am or pm times.

Event	Time
Get up	
Arrive at school	
Lunchtime	
End of school	
Dinnertime	
Bedtime	

3 Use the TV viewing schedule to complete the questions.

Channel A	Channel B	Channel C
News 16:30	*Cricket 16:00*	*Australia Day Special 15:30*
Sport 17:00	*Late News 20:30*	*News 18:00*
Talk Show 18:00		*Movie 18:30*
		Fireworks 22:00

a How long was the cricket on TV? _____

b What time did the fireworks start? _____

c Could Jai watch all the news programs on each channel? _____

d Ella arrived home 15 minutes after the movie started.
What time did she arrive? _____

Unit 26 **Time Problems** (TRB pp. 120–123)
Using units of measurement Use am and pm notation and solve simple time problems
(ACMMG086) (AC)

107

Tenths

$$0.1 = \frac{1}{10} =$$

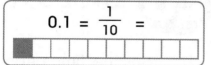

1 Shade the fraction strip to show each amount.

a

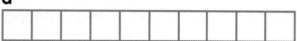

one tenth

b

seven tenths

c

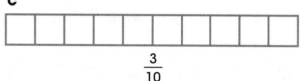

$\frac{3}{10}$

d

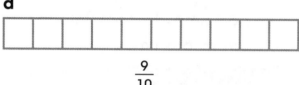

$\frac{9}{10}$

2 Write each fraction as a decimal.

a one tenth _____   **b** nine tenths _____

c three tenths _____   **d** six tenths _____

3 Draw a line matching each fraction to the correct decimal.

a $\frac{2}{10}$ 0.1

b $\frac{7}{10}$ 0.6

c $\frac{1}{10}$ 0.2

d $\frac{6}{10}$ 0.7

4 Write each fraction as a decimal.

a $\frac{3}{10}$ = _____   **b** $\frac{7}{10}$ = _____   **c** $\frac{9}{10}$ = _____

d $\frac{1}{10}$ = _____   **e** $\frac{4}{10}$ = _____   **f** $\frac{5}{10}$ = _____

5 Write each fraction in words.

a $\frac{8}{10}$ = _____   **b** $\frac{2}{10}$ = _____

c $\frac{6}{10}$ = _____   **d** $\frac{10}{10}$ = _____

Extension: Order from **smallest** to **largest**:

nine tenths   0.5 $\frac{8}{10}$ 0.2   one tenth $2\frac{3}{10}$ 5.6   four tenths

108 **Unit 27**

Fractions and Decimals (TRB pp. 124–127)
Fractions and decimals Recognise that the place-value system can be extended to tenths and hundredths. Make connections between fractions and decimal notation
(ACMNA079) **AC**

Hundredths

1 Use fractions to state how much of each 100s grid is shaded.

a  _____

b 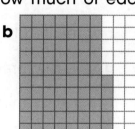 _____

c 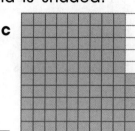 _____

2 Use decimals to state how much of each 100s grid is shaded.

a 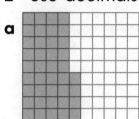 _____

b 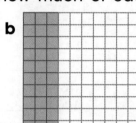 _____

c 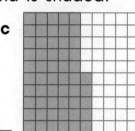 _____

3 Shade part of each 100s grid to match the fraction or decimal.

a $\frac{39}{100}$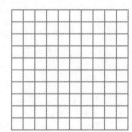

b $\frac{17}{100}$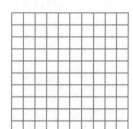

c $\frac{28}{100}$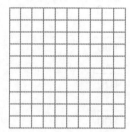

d 0.68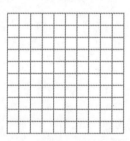

e 0.15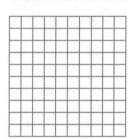

f 0.2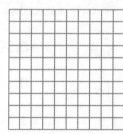

4 Write each fraction as a decimal.

    a five hundredths _____

    b ninety hundredths _____

    c seventeen hundredths _____

    d eighty-four hundredths _____

5 Draw a line matching each fraction to the correct decimal.

0.04	0.4	0.14	0.24
$\frac{14}{100}$	$\frac{40}{100}$	$\frac{24}{100}$	$\frac{4}{100}$

Unit 27 **Fractions and Decimals** (TRB pp. 124–127)
Fractions and decimals Recognise that the place-value system can be extended to tenths and hundredths. Make connections between fractions and decimal notation
(ACMNA079) **AC**

109

Decimals and Fractions

DATE:

1 Write each fraction as a decimal.

    a eight tenths _____

    b three tenths _____

    c forty-one hundredths _____

    d seventy-three hundredths _____

    e one and five tenths _____

    f two and thirty-three hundredths _____

2 Write each fraction in words.

    a $\dfrac{7}{10}$ _____

    b $\dfrac{4}{10}$ _____

    c $\dfrac{29}{100}$ _____

    d $\dfrac{42}{100}$ _____

    e $1\dfrac{1}{10}$ _____

    f $4\dfrac{13}{100}$ _____

3 Draw a line matching each fraction to the correct decimal.

    a $\dfrac{72}{100}$ 0.17

    b $\dfrac{17}{100}$ 0.07

    c $\dfrac{70}{100}$ 0.72

    d $\dfrac{7}{100}$ 0.7

4 Circle the **smaller** number in each pair.

    a $\dfrac{85}{100}$ 0.89   **b** $\dfrac{3}{10}$ 0.1

    c $\dfrac{47}{100}$ 0.68   **d** $\dfrac{5}{10}$ 0.3

    e $1\dfrac{4}{10}$ 1.1   **f** $1\dfrac{17}{100}$ 1.70

5 Write each decimal number or fraction in words.

    a 0.7 _____

    b 0.2 _____

    c 0.22 _____

    d $3\dfrac{29}{100}$ _____

Extension: Draw a diagram to represent 1.47.

Unit **27**
Fractions and Decimals (TRB pp. 124–127)
Fractions and decimals Recognise that the place-value system can be extended to tenths and hundredths. Make connections between fractions and decimal notation
(ACMNA079) **AC**

STUDENT ASSESSMENT

DATE:

1 Write the number shown by each 100s grid as a decimal and a fraction.

a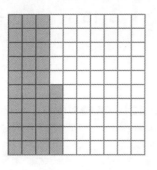

b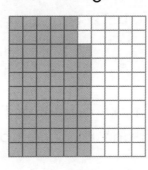

c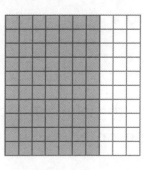

_____   _____   _____

_____   _____   _____

2 Write each number shown as a decimal and a fraction.

a

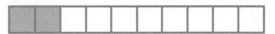

b

_____   _____   _____   _____

3 Draw a line matching each fraction to the correct decimal.

a $\frac{4}{10}$ 0.7   **b** $\frac{2}{100}$ 0.02

$\frac{77}{100}$ 0.4 $\frac{22}{100}$ 1.2

$\frac{2}{10}$ 0.2 $1\frac{2}{10}$ 0.12

$\frac{7}{10}$ 0.77 $\frac{12}{100}$ 0.22

4 Write each decimal or fraction in words.

a 0.6 _____   **b** 0.65 _____

c 1.65 _____   **d** $\frac{7}{10}$ _____

e $\frac{72}{100}$ _____   **f** $3\frac{17}{100}$ _____

Extension: Write a fraction larger than each given number.

a $\frac{42}{100}$ _____   **b** 0.68 _____

c three tenths _____   **d** eighty-nine hundredths _____

Unit
27
Fractions and Decimals (TRB pp. 124–127)
Fractions and decimals Recognise that the place-value system can be extended to tenths and
hundredths. Make connections between fractions and decimal notation
(ACMNA079) **AC**

111

Missing Numbers – Addition

1 Find the missing number in each equation.

a $10 + \underline{\hspace{1cm}} = 15$ **b** $12 + \underline{\hspace{1cm}} = 17$ **c** $15 + \underline{\hspace{1cm}} = 21$

d $\underline{\hspace{1cm}} = 20 + 8$ **e** $\underline{\hspace{1cm}} = 17 + 9$ **f** $\underline{\hspace{1cm}} = 18 + 6$

2 Find the number that will make a total of 55 for each question.
Write an equation. The first one has been done.

a 20 $20 + \underline{\hspace{1cm}} = 55$ $20 + 35 = 55$ missing number is 35

b 30

c 15

d 40

e 27

f 36

3 Find the missing number in each equation.

a $23 + \underline{\hspace{1cm}} = 25 + 30$ **b** $18 + \underline{\hspace{1cm}} = 10 + 10$

c $15 + \underline{\hspace{1cm}} = 22 + 8$ **d** $22 + 24 = 12 + \underline{\hspace{1cm}}$

e $40 + 10 = 35 + \underline{\hspace{1cm}}$ **f** $18 + 14 = 22 + \underline{\hspace{1cm}}$

4 Explain how you worked out Question 3d.

_____

5 Find these numbers.

a When a number is added to 20, the answer is the same as 22 plus 15.

_____

b When a number is added to 35, the answer is the same as 18 plus 48.

_____

Number Sentences (TRB pp. 128–131)
Patterns and algebra Use equivalent number sentences involving addition and
subtraction to find unknown quantities
(ACMNA083) **AC**

Missing Numbers –
Subtraction

1 Find the missing number in each equation.

   a 30 – _____ = 15   **b** 49 – _____ = 17   **c** 45 – _____ = 21

   d _____ = 48 – 37   **e** _____ = 54 – 28   **f** _____ = 61 – 47

2 Find the number that is subtracted from 63 to make each given answer.
Write an equation. The first one has been done.

   a 20   63 – _____ = 20   63 – 43 = 20   missing number is 43

   b 30

   c 13

   d 45

   e 27

   f 36

3 Find the missing number in each equation.

   a 50 – _____ = 60 – 20   **b** 48 – _____ = 55 – 15

   c 38 – _____ = 47 – 28   **d** 44 – 22 = 60 – _____

   e 75 – 54 = 49 – _____   **f** 88 – 66 = 38 – _____

4 Explain how you worked out Question 3e.

_____

5 Find these numbers.

   a When a number is subtracted from 70, the answer is the same as

      90 minus 50. _____

   b When a number is subtracted from 35, the answer is the same as

      57 minus 43. _____

Unit 28 **Number Sentences** (TRB pp. 128–131)
Patterns and algebra Use equivalent number sentences involving addition and
subtraction to find unknown quantities
(ACMNA083) **AC**

113

I Am Thinking ...

1 I am thinking of a number:

   a I add 20, then subtract 12. My new number is 62.
     What was my starting number?

   b I subtract 8 and then add 12. My new number is 37.
     What was my starting number?

   c I add 15, and then subtract 22. My new number is 48.
     What was my starting number?

2 Find these numbers.

   a When a number is added to 25, the answer is the same as 67 minus 19.

   b When a number is subtracted from 100, the answer is the same as
     46 plus 17.

   c When a number is added to 120, the answer is the same as
     200 minus 63.

3 Find the missing number in each equation.

   a $100 + \underline{\hspace{1.5em}} = 193 - 45$ **b** $152 + \underline{\hspace{1.5em}} = 200 - 13$

   c $125 + 126 = 300 - \underline{\hspace{1.5em}}$ **d** $167 + 194 = 502 - \underline{\hspace{1.5em}}$

114 **Unit 28** **Number Sentences** (TRB pp. 128–131)
Patterns and algebra Use equivalent number sentences involving addition and
subtraction to find unknown quantities
(ACMNA083) **AC**

STUDENT ASSESSMENT

1 Albert is thinking of a number. He adds 17 and then subtracts 9. His new number is 132. What was his starting number?

2 Explain how you worked out the answer to Question 1.

_____

_____

3 When a number is added to 33, the answer is the same as 87 minus 39. What is the number?

4 Explain how you worked out the answer to Question 3.

_____

_____

5 Find the missing number in each equation.

    a 112 + _____ = 200 − 74   **b** _____ = 161 − 59

    c 140 + 10 = 85 + _____   **d** 210 − _____ = 140 + 60

Number Sentences (TRB pp. 128–131)
Patterns and algebra Use equivalent number sentences involving addition and subtraction to find unknown quantities
(ACMNA083) **AC**

115

Picture Graphs

You will need: a ruler, coloured pencils or felt pens

1 Using the data in the table below, create a picture graph on the grid paper.

Colour of marble	Number in the bag
red	5
blue	3
green	6
yellow	1
pink	2
orange	2
brown	0

2 Using the data in the table below, create a picture graph on the grid. Choose suitable pictures, and also have 1 picture = 10 days.

Type of weather	Number of days
sunny	44
cloudy (but no rain)	14
rainy	4

3 On a sheet of paper, write 3 sentences about each of your graphs.

Displaying Data (TRB pp. 132–135)
Data representation and interpretation Construct suitable data displays, with and without the use of digital technologies, from given or collected data.

Include tables, column graphs and picture graphs where one picture can represent many data values
(ACMSP096) **AC**

Collecting Data

One way to collect data is using Venn diagrams and 2-way tables.

This is a survey about skateboards and scooters.

1 Survey your classmates to find out how many have a skateboard, a scooter, both or neither. Complete the table below.

	Yes
Do you have only a skateboard?	
Do you have only a scooter?	
Do you have a skateboard and a scooter?	
Do you have neither a skateboard nor a scooter?	

2 Place this data into a Venn diagram.

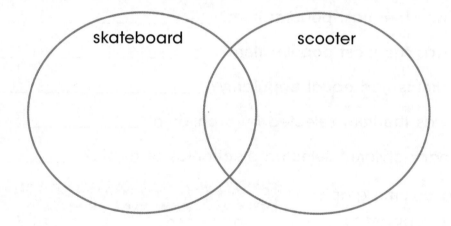

skateboard   scooter

3 What does the overlapping circle represent?

_____

4 Display the data you have collected in a 2-way table.

	Has a scooter	Doesn't have a scooter
Has a skateboard		
Doesn't have a skateboard		

5 What do you notice about the Venn diagram and the 2-way table?

_____

Unit 29 **Displaying Data** (TRB pp. 132–135)
Data representation and interpretation Construct suitable data displays, with and without the use of digital technologies, from given or collected data.

Include tables, column graphs and picture graphs where one picture can represent many data values
(ACMSP096) **AC**

117

Column Graphs

1 75 children were asked what the number 1 item on their birthday list was. Here is a graph showing their responses.

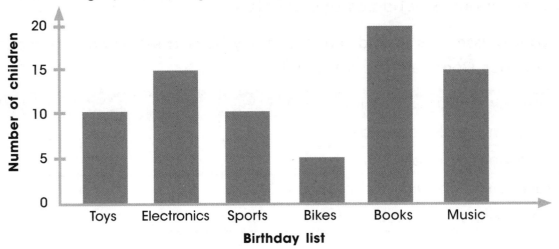

Use the graph to answer the questions.

a What was the most popular item? _____

b What was the least popular item? _____

c Which items had equal popularity? _____

d What was the item selected by 5 children?_____

e How many children selected electronics or music? _____

2 Create a column graph on the grid paper to show the number of animals at the farm.

Sheep	Cows	Chickens	Horses	Ducks	Goats
50	100	75	75	65	55

3 On a sheet of paper, write 3 statements about the graph.

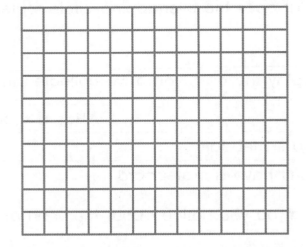

Displaying Data (TRB pp. 132–135)
Data representation and interpretation Construct suitable data displays, with and without the use of digital technologies, from given or collected data.

Include tables, column graphs and picture graphs where one picture can represent many data values
(ACMSP096) **AC**

STUDENT ASSESSMENT

1 Look at the graph showing the forms of transportation used by different numbers of people. Use the graph to answer the questions.

Main forms of transport

| 1 picture = 10 people |

Number of people

car   bus   train   tram   motorbike   pushbike

Transport

a What was the most popular form of transport? _____

b How many people mainly used trains? _____

c Which mode of transport was selected by 10 people? _____

d What was the total number of people who selected buses and trams? _____

2 Using the grid paper, convert the picture graph into a column graph.

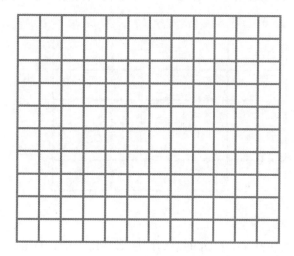

Unit
29

Displaying Data (TRB pp. 132–135)
Data representation and interpretation Construct
suitable data displays, with and without the use of digital
technologies, from given or collected data.

Include tables, column graphs and picture graphs where
one picture can represent many data values
(ACMSP096) **AC**

119

Graphing the Weather

Look at the data about the weather in the table.

Day	Week 1	Week 2	Week 3	Week 4
Monday	☀	☁	☁	☀
Tuesday	☁	☁	☁	☀
Wednesday	☁	☀	☁	☀
Thursday	☀	☀	🌧	☀
Friday	☁	☀	🌧	☁
Saturday	🌧	☁	☁	☁
Sunday	🌧	☁	☀	🌧

1 Complete the table with the weather data shown in the table above.

Type of weather	Number of days (tally)	Total
Sunny		
Cloudy		
Rainy		
Foggy		

2 Create a column graph of the weather data.

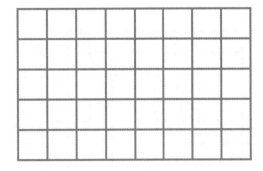

3 On a sheet of paper, write 3 statements about the graph.

Interpreting Data (TRB pp. 136–139)
Data representation and interpretation Evaluate the effectiveness of different displays in illustrating data features including variability
(ACMSP097) **AC**

Interpreting Data

Look at the data about Kay's crops.

Crop	1 picture = 10	Total number
Lettuce		
Carrots		
Potatoes		
Sweet corn		
Broccoli		
Pumpkins		

1 Complete the table above with the total number of each vegetable.

2 Find the total of all Kay's crops. _____

3 Create a column graph showing Kay's crops. Hint: there should be 50 lettuces.

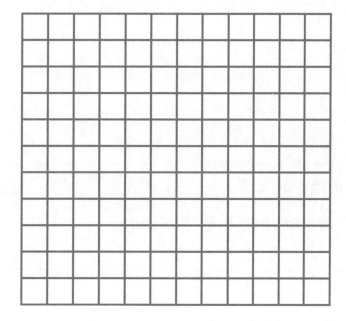

Unit 30 **Interpreting Data** (TRB pp. 136–139)
Data representation and interpretation Evaluate the effectiveness of different displays in illustrating data features including variability
(ACMSP097) **AC**

121

At the Beach

Look at the graph about the beach.

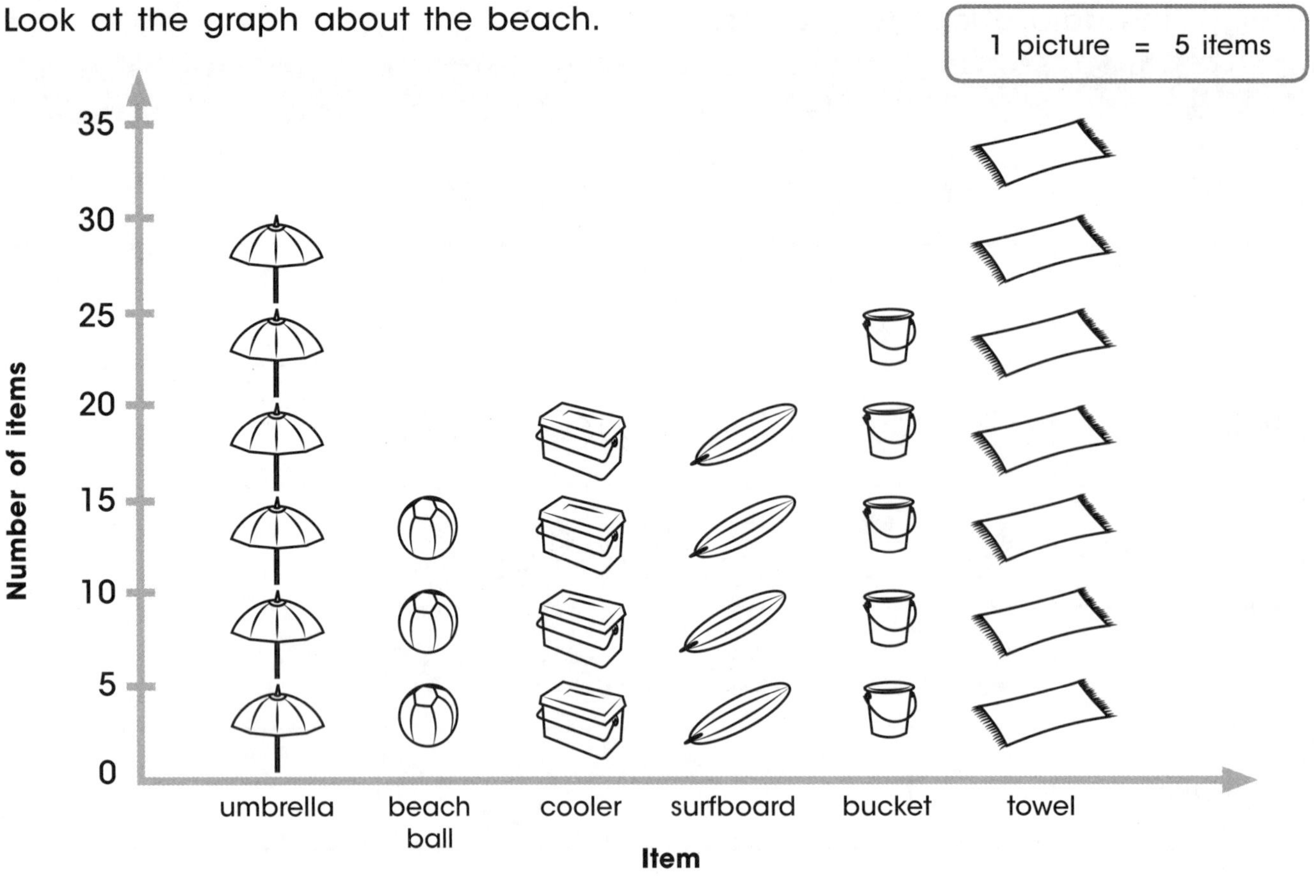

1 picture = 5 items

1 Create a table that shows the data from the graph.

2 Write a heading for the data. _____

3 Write 3 facts that can be seen from the data.

_____

_____

Interpreting Data (TRB pp. 136–139)
Data representation and interpretation Evaluate the effectiveness of different displays in illustrating data features including variability
(ACMSP097) AC

DATE:

Look at the graph of beach animals.

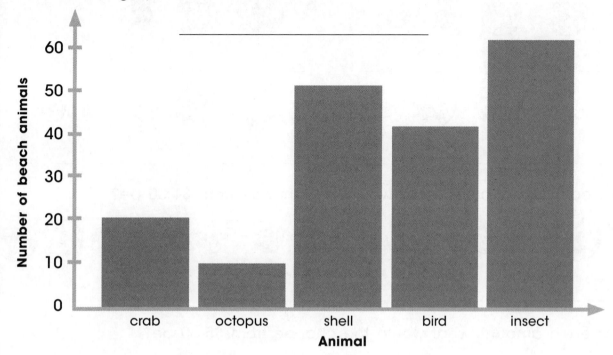

1 Give the graph a heading.

2 List 3 questions that could be asked about the data.

a _____

b _____

c _____

3 Make the information from the graph into a table of data.

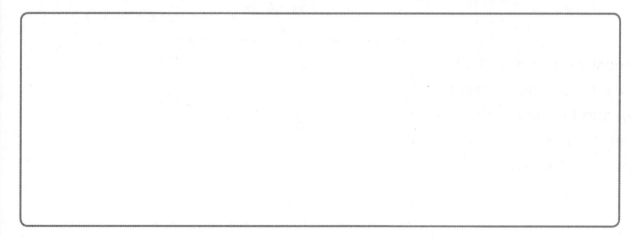

Unit
30 **Interpreting Data** (TRB pp. 136–139)
Data representation and interpretation Evaluate the effectiveness of different
displays in illustrating data features including variability
(ACMSP097) **AC**

123

Change

1 What is the value of each set of coins?

a   **b**

_____   _____

c   **d**

_____   _____

2 For each set of coins, what would the change from $1.00 be?

a   **b**   **c**

_____   _____   _____

3 For each amount, what would the change from $5.00 be?

a $3.50 _____   **b** $2.90 _____

c $4.10 _____   **d** $1.75 _____

4 Explain how you found the answer to Question 3d.

_____

5 Round the amounts and then find the change from the **nearest** dollar.

a $1.94 _____   **b** $2.61 _____

c $4.12 _____   **d** $9.98 _____

Extension: Make $10.00
in 4 different ways, using
any combinations of
notes and coins.
Draw the 4 ways.

124 **Unit 31** **Money** (TRB pp. 140–143)
Money and financial mathematics Solve problems involving purchases and the calculation
of change to the nearest five cents with and without digital technologies
(ACMNA080) **AC**

Off to the Movies

1 You are off to the movies and lunch with your friends. Circle your choice of movie.

2 You decide to have a snack at the movies. Write what you select.

Snack menu	
popcorn	$5.00
drink	$3.00
ice-cream	$3.50
chocolate bar	$2.50
packet of chips	$3.00

I selected:

3 What is the total cost of going to the movies? _____

4 Write what you select for lunch.

Lunch menu	
pie	$5.00
pastie	$5.00
sandwich	$6.00
chips and gravy	$4.00
drink	$3.00
biscuit	$1.50
cake	$3.00

I selected:

5 What is the total cost of lunch? _____

6 What is the total cost of going to the movies + lunch? _____

7 Your mum gave you $50.00. How much change do you give to your mum?

_____

Unit **31** Money (TRB pp. 140–143)
Money and financial mathematics Solve problems involving purchases and the calculation of change to the nearest five cents with and without digital technologies
(ACMNA080) **AC**

125

In China

Dharma is visiting China.

In China, the base unit of money is the Yuan.

In China, there is a 1 Jiao coin and a 5 Jiao coin. 1 Yuan = 10 Jiao. There is both a 1 Yuan coin and a 1 Yuan note. There is also a 5 Yuan note, a 10 Yuan note, a 20 Yuan note, a 50 Yuan note and a 100 Yuan note.

> Here are samples of the money:

1 Dharma bought some lunch of rice and vegetables and a drink. The cost was 5 Yuan. What combinations of money could Dharma use?

_____

2 Dharma went to a theme park for the day. The cost was 65 Yuan. What change would Dharma receive from 100 Yuan?

_____

3 What do you notice that's different about Chinese money compared to the money in your country? What is the same?

_____

Extension: Can you find out the value of 1 Yuan in Australia? What is 100 Yuan worth?

STUDENT ASSESSMENT

DATE:

You will need: a calculator

1 What is the value of each set of coins and/or notes?

a   _____

b   _____

c    _____

d    _____

2 For each item, what would the change from $10.00 be?

a  $7.50   **b** $3.25   **c** $4.50   **d** | **Gift Card** | $6.95 |

_____   _____   _____   _____

3 Look at the items for sale.

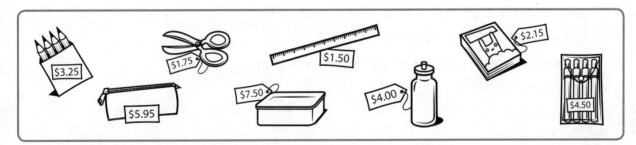

$3.25 $1.75 $1.50 $2.15 $5.95 $7.50 $4.00 $4.50

a Select 3 items and find their total cost (using a calculator if required). _____

b How much change will you get if you pay for the 3 items with $20.00? _____

Unit
31
Money (TRB pp. 140–143)
Money and financial mathematics Solve problems involving purchases and the calculation
of change to the nearest five cents with and without digital technologies
(ACMNA080) **AC**

127

Word Problems

Write a number sentence to solve each word problem.

Word problem	Number sentence and answer
1 Tom had 5 soft toys, 3 dolls and 6 trains. How many toys did he have altogether?	
2 Kayla had 3 golf balls, a soccer ball, 2 footballs and 4 tennis balls. How many balls did she have in total?	
3 For the school fete, 4 HS had collected 215 plants and 76 pots to sell at their garden stall. How many items did they have altogether?	
4 The shop had 75 ice-creams at the start of the day. They sold 27. How many were left?	
5 There were 800 cards in a set. Noah had collected 665 of them. How many more did he need to complete the set?	
6 There were 12 biscuits in a packet. Chloe gave 3 to her brother, 2 to her little sister and 1 to her dad. How many biscuits were left?	
Challenge! The canteen had 50 pies for lunch orders. The Year 1–3 students bought 15 pies and the Year 4–6 students bought 12 pies. How many were left?	

Word Problems (TRB pp. 144–147)
Patterns and algebra Solve word problems by using number sentences involving multiplication or division where there is no remainder
(ACMNA082) **AC**

Multiplication Problems

Complete the gaps in the table by writing the number sentence and answer, or a word problem.

Word problem	Number sentence and answer
1 On each day of the week, Nana received 8 letters. How many letters did she receive for the week?	
2	$6 \times 12 =$
3 On the farm there were 200 horses and 100 emus. How many legs were there altogether?	
4	$90 \div 10 =$
5 Jo carefully shared his 48 strawberries among himself and his 5 friends. How many strawberries did each person receive?	
6	$36 \div 6 =$
Challenge! Yesterday, I put some counters into groups with the same number of counters in each group. There were 24 counters in total, but I cannot remember the number of groups. What might my groups have been?	

Unit 32 **Word Problems** (TRB pp. 144–147)
Patterns and algebra Solve word problems by using number sentences involving multiplication or division where there is no remainder
(ACMNA082) **AC**

129

Division Word Problems

Complete the table by:

- including the related multiplication equations.

- writing a word problem for each equation.

	Number sentence	Multiplication equation	Word problem
1	18 ÷ 3 =	6 × 3 = 18 Or 3 × 6 = 18	
2	42 ÷ 6 =		
3	50 ÷ 10 =		
4	36 ÷ 4 =		
5	40 ÷ 5 =		

Word Problems (TRB pp. 144–147)
Patterns and algebra Solve word problems by using number sentences involving
multiplication or division where there is no remainder
(ACMNA082) AC

Unit
32 STUDENT ASSESSMENT

You will need: a calculator

1 Write each problem as a number sentence and solve.

a Annabel had 12 pencils, 12 felt pens, 2 pens, an eraser, a ruler and a pair of scissors in her pencil case. How many items did Annabel have in her pencil case?

b Cody was asked to sort 30 pencils into 5 containers. How many pencils were in each container?

2 Complete the think board.

Answer: 9	Equation:
Diagram of your equation:	Word problem for your equation:

Unit
32 **Word Problems** (TRB pp. 144–147)
Patterns and algebra Solve word problems by using number sentences involving multiplication or division where there is no remainder
(ACMNA082) **AC**

131

Symmetrical Numbers and Shapes

You will need: BLM 17 '1 cm Grid Paper'

List the digits 0–9 in the spaces below, writing them so that they fill the space.

<table>
<tr><td></td><td></td><td></td><td></td><td></td></tr>
<tr><td></td><td></td><td></td><td></td><td></td></tr>
</table>

1 Draw lines of symmetry on the numbers that have them, using dotted lines.

2 Create some 4-digit numbers in which all digits are symmetrical.

3 It is possible to find symmetrical patterns by shading 4 squares on a 3 × 3 grid. For example (right):

Using grid paper, see how many solutions you can find. Don't forget to draw in the line of symmetry.

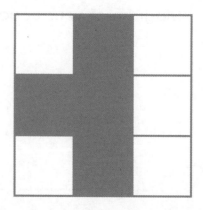

Patterns (TRB pp. 148–151)
Location and transformation Create symmetrical patterns, pictures and shapes with and without digital technologies
(ACMMG091) **AC**

Describing Art

Look at the following artwork examples.

Next to each artwork, write a description of the art and the patterns you see, including elements of symmetry as well as the shapes used in the design.

Extension: Look in magazines for patterns that contain elements of symmetry. Collect the images and create a poster.

Unit 33

Patterns (TRB pp. 148–151)
Location and transformation Create symmetrical patterns, pictures and shapes with and without digital technologies
(ACMMG091) AC

133

Not Patterns

Using the grid below, the challenge is to draw as many different rectangles as possible that are **not** the same shape or same size. So in this exercise the aim is **not** to make a pattern.

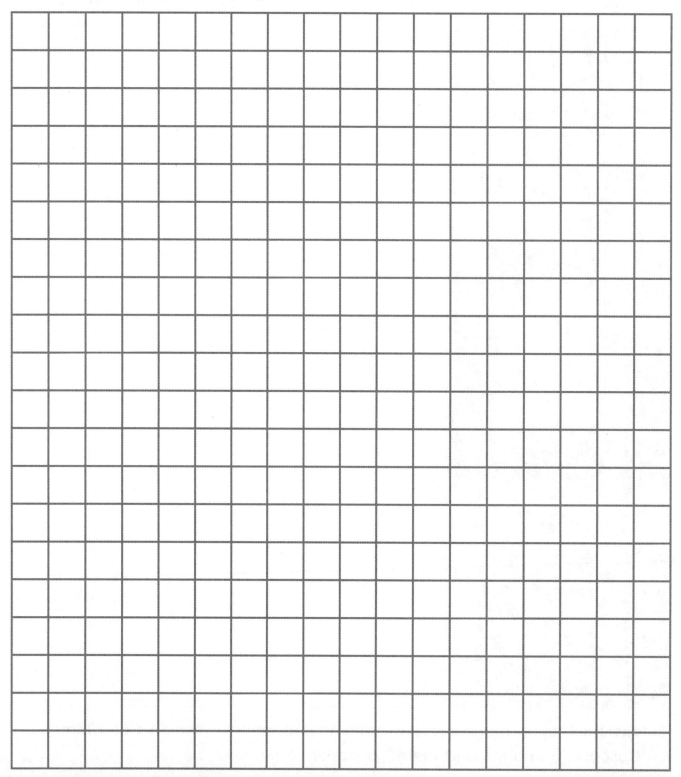

Patterns (TRB pp. 148–151)
Location and transformation Create symmetrical patterns, pictures and shapes with and without digital technologies
(ACMMG091) AC

STUDENT ASSESSMENT

1 Look at each picture and complete the other half around the line of symmetry.

a    **b**    **c**

2 Create a pattern using the grid.

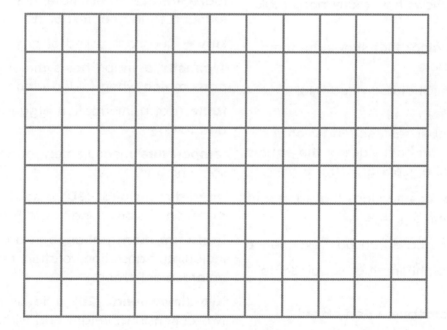

3 Describe your pattern above. Make sure you name the shapes you used. _____

_____

4 Does your pattern have symmetry? Explain. _____

_____

Unit
33

Patterns (TRB pp. 148–151)
Location and transformation Create symmetrical patterns, pictures and
shapes with and without digital technologies
(ACMMG091) **AC**

135

Maths Glossary

acute angle an angle less than 90°

analogue clock a clock with two hands and 12 numerals

angle the space between the intersection of two straight lines

axes the lines that form the framework of a graph, i.e. x-axis and y-axis

BODMAS gives the order for solving equations; it stands for: **B**rackets, **O**f, **D**ivision, **M**ultiplication, **A**ddition, **S**ubtraction

capacity how much a container can hold

chance the likelihood something will happen

data factual information gathered for research

denominator the bottom number in a fraction, which shows how many parts make up a whole

digital clock a clock that shows the time using only numbers

divisor a number that is divided into another number

equation a number sentence in which the numerical value on both sides of the equals sign is the same, e.g. $6 \times 5 = 10 \times 3$

equivalent fractions fractions that represent the same amount, e.g. $\frac{1}{2} = \frac{2}{4} = \frac{3}{6} = \frac{4}{8}$

estimate a guess based on past experience

horizontal a straight line that is parallel to the horizon

mass quantity of matter in an object

mixed number a number that consists of a whole number and a fraction

number line a line marked with numbers to show operations or patterns

number sequence an ordered set of numbers

numerator the top number in a fraction, which shows how many parts of the whole there are

obtuse angle an angle greater than 90°

parallel lines a set of lines that remain the same distance apart and do not intersect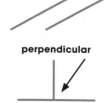

perpendicular line a vertical line that makes a right angle where it meets a horizontal line

place value the value of a digit dependent on its position in a number

polygon a plane shape with three or more sides

probability the chance a particular outcome will occur compared to all outcomes

reflex angle an angle between 180° and 360°

right angle an angle of exactly 90°

rounding to change the value of a number to make it easier to estimate; numbers can be rounded to the nearest 5, 10, 100, etc.

survey questions asked of a group of people

symmetry a shape has symmetry if both its parts match when folded along a line

table (data) information organised in columns and rows

temperature measurement of how hot or cold something is

three-dimensional (3D) a solid shape with three dimensions: length, width and height

transformation a change in position or size including: translation, rotation, reflection or enlargement (zoom)

two-dimensional (2D) a flat shape with only two dimensions: length and width

Venn diagram overlapping circles used to show different sets of information

vertical a straight line at right angles to a horizontal

volume the space occupied by a 3D object

weight mass affected by gravity